Fearless, Sleepless, Deathless

GREAT CIRCLE BOOKS

Great Circle Books publishes literary nonfiction—including memoir, literary journalism, personal and lyric essays, and works that defy easy classification—by emerging writers. The Great Circle is the intersection of lines on the celestial sphere—lines overlapping and creating new entryways of understanding. In that spirit, Great Circle Books seeks, through innovative work, to merge the human experience with our relationship to place.

A complete list of books published in Great Circle Books is available at https://uncpress.org/series/great-circle-books.

Fearless, Sleepless, Deathless

What Fungi Taught Me about Nourishment, Poison, Ecology, Hidden Histories, Zombies, and Black Survival

Maria Pinto

The University of North Carolina Press
CHAPEL HILL

This book was published with the assistance of the Wells Fargo Fund for Excellence of the University of North Carolina Press.

Designed by Lindsay Starr
Set in Iowan Old Style by codeMantra
Manufactured in the United States of America

Cover art: Mushrooms © Adobe Stock / New Africa.

Complete Library of Congress Cataloging-in-Publication Data is available at http://lccn.loc.gov/2025025227.

ISBN 9781469689791 (paper; alk. paper)
ISBN 9781469689807 (epub)
ISBN 9781469689814 (pdf)

For product safety concerns under the European Union's General Product Safety Regulation (EU GPSR), please contact gpsr@mare-nostrum.co.uk or write to the University of North Carolina Press and Mare Nostrum Group B.V., Mauritskade 21D, 1091 GC Amsterdam, The Netherlands.

For that bright fly agaric who turned my head. For those already in love with fungi and for those who wish to apprentice themselves to Anansi, to never feel the urge to squish his emissaries in the networked corners of their homes again.

Nature, through all her kingdoms, insures herself. Nobody cares for planting the poor fungus: so she shakes down from the gills of one agaric countless spores, any one of which, being preserved, transmits new billions of spores tomorrow or next day. . . . A fearless, sleepless, deathless progeny, which is not exposed to the accidents of the weary kingdom of time.

—Ralph Waldo Emerson, "The Poet"

To fungi, our world of light and air is their underland, into which they tentatively ascend here and there, now and then.

—Robert Macfarlane, *Underland*

We are the metaphors for them.

—Anne Boyer on humans and mushrooms

CONTENTS

ACKNOWLEDGMENTS

This book would not exist without the many thriving mushroom communities, online and IRL, and everyone in them who helped dig the rabbit hole. There would be nothing in your hands if not for the efforts of Arielle, Almah and Treva, my loves at Heauxfort, the BADhouse Salon crew, Bobby, Emily, Jasmine, Lindsay, Tyler, Melinda, Christine, Lauren, Gordon, Justin, Sidi, João, Matt, Blacki, Sneha, Sigrid, Carlos, Mario, Belen and Sil, Kieran, Billy and Bill, my Dreyfoos crew, Theresa and Teresa, Eson, Kara, and Faith Smith. Thanks to all who make the residencies at the Mastheads, MASS MoCa, Vermont Studio Center, and Oak Spring Garden Foundation happen, and those I met in these places who shared their work, urged me on, and with whom I shared walks in the woods. Thank you, Mums and Jermaine and Akeem, and everyone back home who has put up with or encouraged this obsession. Thank you, Lucas, thank you, Zoe, for seeing, sparking, believing, nudging, shaping, advocating. Thank you, Walter, for knowing and reading and loving me best.

Fearless, Sleepless, Deathless

Introduction

Picture a living weave, an omnipresent blooming web, questing alongside and inside much of life on earth, undergirding it, nourishing it, digesting and decomposing it, linking it, mimicking it, wearing it, decorating it, sickening and humbling it, defending it, amusing and amazing it, eating its shit. This is the mycelium, the mass of threads that make up most fungi. The individual filaments, known as hyphae, collectively constitute much of the kingdom that labors at some of the least perceptible, yet most intricately indispensable, jobs on the planet. Their toil is often invisible, microscopic, underground, inside our guts and on our skin, beyond detection until it is spectacularly *not* invisible—breaking forth into our upper world as fruit bodies, the mushrooms our mothers warned us not to eat.

Before mycelium attends to any of its* vital earthly vocations, it generally starts its life as a spore. Spores are light and small and everywhere, so morphologically diverse that their shapes, under a microscope, are used to identify the fruit bodies from which they spring. Like seeds, these propagules are built to travel. For the spores of some terrestrial fungi, however, escaping the layer of still air right next to the ground requires ingenuity. Bird's nest fungus spores (which look like, you guessed it, bird's nests in miniature, down to the egg-shaped discs at their centers) are splashed into the air by raindrops. Subterranean truffles use their intoxicating perfume to seduce deer into digging for and devouring them, the spores eventually hitching a ride to a new location via ungulate digestive tracts. An artist's conk, growing shelf-like from the side of a standing dead tree, relies on the wind to disseminate its spores and might produce 350,000 of them a second for six months of the year, in hopes that at least one will go the distance.[1] Indeed, the fungus called wheat rust has been recorded

* In speech, I'm in the process of replacing all instances of "it" to refer to anyone living—an oak, a horse, a tardigrade—with the same pronoun I would accord a fellow *Homo sapiens*. While "they" flows easily enough out loud, it's trickier on the page, and I'm still looking for a graceful way to heal my language on the page. In the meantime, please pardon my dust.

hitching a 1,243-mile ride on what must have been a very wicked wind indeed.[2] Spores ride the waters and sail on winds in search of new places to flourish, and once they've found the perfect conditions in which to germinate (which can occur after many years, freezings-over, and droughts) those spores beget hyphae, which beget mycelium, which is, as biologist Merlin Sheldrake beautifully puts it, "living, growing, opportunistic investigation—speculation in bodily form."[3]

Without this process, by which fungi weave life and the world into being, the very earth as we know it, the soil that makes our existence possible, would be a wild dream whispered weakly from plant to plant. It is through the speculation—through the questing and devouring, the digesting and engendering, the dismantling and devising—that fungi build and sustain our world. Unlike plants, who produce their own food through photosynthesis, or animals, who feed by putting food into their bodies, fungi *myceliate* their food source, sending enzymes out to process the materials they move through. While you take food from the outside in, fungi digest matter externally, in the process releasing nutrients for others into the world.

And without members of this understudied kingdom, that crucial moment hundreds of millions of years ago when plants colonized the land might still just be a potentiality. I have found it nearly irresistible to humble myself before the what-ifs of the fungal world. What if these earth-building fungi had never committed to turning stone into soil? What if mycelia never stitched themselves into the roots of plants, helping those plants make use of the nutrients in that soil? What if mycelia never stitched themselves into our animal bodies, as numerous as the years through which we have adapted to live, fungus and animal, mutualistically?

Mycelium built and then connected our world.

One way to think about the anatomy of this book, then, is that its essays will imitate the mycelium, as often as possible initiating an investigation without a preconceived goal in mind. Their shape will be indeterminate: as speculative instruments they'll contract, explore, and expand as necessary. To *essay*—literally "to try"—is, in other words, to myceliate.

Of the estimated 2–3 million species of fungi animating the earth, only about 155,000 have been described by Western science. That leaves about 92 to 95 percent that occur unnamed by us.[4] If fungi are so ubiquitous and important, so fascinating and strange, why do we know so little about them? An easy answer is that even when we have visible proofs of their microscopic, interstitial, underground existence in the form of mushrooms and other fruit bodies, those proofs can be fleeting. Their rhythms are their own. Crowded troupes of ink caps (*Coprinellus* sp.) muscle up from the depths overnight at the bases of trees and on stumps only to deliquesce—that is, to turn into a mass of black goo—in a matter of twenty-four hours. I've noted stinkhorn "eggs" in parking lot mulch near a woodland in the morning as I've set out on a walk; later that evening found their pale stalks and green heads emerged and standing proud six inches above the ground; then come back a couple days later to find the ground where they stood bare of any fungal traces.

In much the same way that anything brought into a lab setting behaves differently than it does in the wild, fungi defy cloistered forms of study. It's hard to examine a thing's stitching, and the things it stitches together, by unraveling the stitch. To complicate matters, when we* as naturalists and scientists have paid attention to the ways fungi move through the world, we have often done so in a phytocentric (plant-focused), or a zoocentric (animal-focused) way: thinking only about what fungi will do to the trees and crops that we live with and use, noticing them most when our pets contract ringworm or our wheat and rice crops

* The "we" of this book will shift according to the author's expanding and contracting affinities—we as humans, we as North Americans, we as Black people, we as Westerners, we as naturalists and scientists (as in the passage above), we the desirous of collective liberation, and on and on—referring to these identities as needed.

contract *Magnaporthe oryzae*. Even now, during a flourishing of fungal study, we've still only named fewer than 10 percent of the millions of species believed to exist.

Rarely have we studied fungi on their own terms. Consider that ecologist Robert Whittaker, the first Westerner to propose funga as a taxonomical kingdom distinct from flora and fauna, did so as recently as 1969.[5] Mycophile Rachel Zoller points out that we were landing on the moon just as we recognized the queendom as its own realm, which means our longstanding innocence of fungal ways isn't due to a lack of exploratory ingenuity.[6]

Fungi have made it possible to think with a different model: one in which the aboveground is just some of the story, and perhaps not the most interesting part. Indeed, the aboveground can be made just as alien as its more hidden neighbor, when viewed through a fungal lens. When I look at the mysteries and illegibilities of fungi, I can relate. Learning about them has taught me to dwell more comfortably in mystery, even as I pursue answers and better questions. Plenty of the material conditions and habits of thought that root and connect people to things and to one another on this planet can feel somewhat notional for me, as a neurodivergent, naturalized US citizen from the Caribbean, and a descendant of enslaved people. I've lived my life within an epistemological framework in which those true, to-the-center-of-the-earth roots were for others, not me. And since my family tree came from a clear-cut forest, I would of necessity be relegated to a surface kind of existence. But the more I learn about the ancient origins and tantalizing futurity of fungi, about their centrality to healthy ecosystems and their adaptability, about their potential for earthly and mental remediation, the more I've realized that my kinship lines feel more mycelial than tree-like. That, like fungi, the stuff I'm made of (that we're all made of) has the power to move in darkness, to thrive undetected, to quietly work until such a time as there's nothing left to do but fruit. All eukaryotes have a common ancestor if we go back 2 billion years, anyhow. In this way, I recognize fungi as my kin. I truly can, and do, relate.

As of this writing, huge strides toward an ecologically sound understanding of our planet are being made by using a fungal lens. The North American Mycoflora Project, which turned into the Fungal Diversity Survey (FunDiS), is mounting a heroic effort to encourage citizen scientists in their documentation of the rich fungal biome we live with every day. The scientists behind SPUN (Society for the Protection of Underground Networks), an international effort to map our mycelial networks by taking thousands of richly myceliated soil samples from around the world, began work in 2022.[7] Popular participation in science is enjoying a nineteenth-century-style uptick: via online applications like iNaturalist and Mushroom Observer, crowdsourcing is expanding the mycological record. At the same time, those who've thought of themselves as the center of humanity are being forced to admit that the margins (and those they sought, and seek, to bury underground there) are where the keys to our continued existence can be found. Reckonings abound as our planet continually tips toward apocalypse. And as this dark renaissance takes root, I've begun using a fungal lens to bring my own mental and ancestral underground into focus. The essays here are a record of that exploration. As I sought a path to a more integrated sense of my past and the future, I found there's nothing like the study of fungi to encourage ecological thinking and explode notions of discrete individuals, lone actors, and single-stream histories.

IN SEARCH OF

Junjo

(THE JAMAICAN MUSHROOM)

On dirt, Anansi, heartbreak,
big box parking lots, the
Junjo–djon djon connection,
the clan that counts them
as kin, what happens to
an identity in the woods,
and what it means to love
Northeastern temperate forests

Those who cyan' hear must feel.

—Jamaican proverb

In a grassy island of a Target parking lot, I spot a flush of pleasantly plump, rose-red-capped mushrooms and make a beeline for them, my mother close behind. I drop to my knees, take note of the live oaks nearby, and tap the mushroom domes, pleased to hear the satisfying *thunk* that means they're solid inside, not squirming with maggots.

"Aren't they beautiful?" I ask, snapping some photos before pulling out my fancy French mushroom knife to cut six or so fruit bodies off at the feet. My mother says yes unironically, which surprises me until I try to see them through her eyes—this is likely the first time she's ever encountered such zesty fungal morphology up close. I stand and score the yellow pore surface of one of the mushrooms with my knife.

A family walks by, bags in hand. The little kids pause their bickering caravan to rubberneck, gaping at me, my knife, and my mother.

The wound turns color instantaneously.

"That quick blue bruising is a hint about who she is," I say, watching the family pile into their city-block-sized SUV. My mother indulges me with an "okaaay," and unlocks the car so I can deposit the mushrooms on the floorboard. As we shop for face cream and a whisk, I think about how the search for mushrooms has rendered the land I used to tread inattentively in three dimensions. There's no dead space anymore—it's all animate. Last week I'd crouched to examine a train wrecker mushroom growing from a downed fence post in an abandoned lot and was reminded of the concept of edgelands, as described by poets Paul Farley and Michael Symmons Roberts: locations in urban and suburban environments that are "so difficult to acknowledge they barely exist."[1] But of course they, too, constitute the Earth. Later, in the lawn in front of my mother's condo, I see more rose-red boletes. I grab them and point out to her that these, unlike the Target specimens, seem to be wearing fishnets on their stems.

While my mother trains one eye on her fifth consecutive episode of the '90s British drama *Heartbeat* and the other eye on a shopping tab in her browser, I sprawl on the floor with *Boletes of Eastern North America*, a laptop open to mushroomobserver.com, and one of my rose-capped, blue-staining boletes lying on the page next to its two-dimensional mirror image. So far I've arrived at *Pulchroboletus rubricitrinus* for the parking lot specimens and *Exsudoporus floridanus* for the lawn variety. Keying them out took a minute, and throughout the process my mother kept asking if I knew what they were yet—an annoying karmic comeuppance for my own endless questions during long road trips, three decades before.

Now I say, "You see all these guys? How they look almost identical at first glance, but they're actually completely different species?" I flip through the pages to punctuate my point.

"Those are all different?" She gapes, her glasses way down on the bridge of her nose.

"Yep—some are choice edibles, some reportedly taste bad, and some will put you on the toilet for a while."

My mother shakes her head and looks back at the television.

"In Jamaica we only had two names for mushrooms. Junjo, and duppy umbrella."

I pause mid-flip, processing what she just said.

"What?"

"Those were the two categories. It was much simpler."

"Duppy" of course means "ghost," so "duppy umbrella" is a word picture akin to "toadstool," but the other term had tripped my memory. "Junjo," its syllables pleasingly reverberant in the mouth, sent me spinning down a Google rabbit hole, as might any offhand comment linking one's obsession to one's culture in a previously undreamt-of way. This hole in the grass was quite welcome, not least because I was slated to give a talk about fungi and the African diaspora in a couple of months and had been hunting for ways to braid my own Jamaican American identity into the conversation. I'd hit a wall there. Though I had a hunch that, like so much about my people, their relationships to foraging had been coded, forcibly hidden, buried, or never written about, still everything I had managed to find suggested that Anglophone Caribbeans just wanted nothing to do with mushrooms. To the point where they just weren't discussed.

Nevertheless, I'd heard "junjo" before. When I was growing up, my family used the Patois word to refer to dirt or schmutz that had collected in a place, giving it an untidy appearance. This made sense because the word's more specialized usage referred to fungi that present as a fine coating on something—molds and mildews. It was a letdown if the only time Jamaicans used their most common word for fungi, they were doing so to disapprove—say, to tell a kid who had just polished off a mudpie to wash the filthy junjo off her face. But somehow I felt the word held more.

I said the word—"junjo, junjo, junjo"—repeatedly, there in the living room while my mother tried to follow a conversation between the cute constable and his love interest, until it hit me. (Not my mother, though I'm sure she wanted to.)

"Huh. Junjo kinda sounds like 'djon djon.'"

"What?" my mother asked, distracted.

"Nothing."

"Djon djon" is the Haitian Kreyól word for both "mushroom" and the specific type of mushroom that Haitians use to flavor, and dye black, one of their national dishes, *diri ak djon djon*. Rice cooked in the liquid left from boiling a dried *Psathyrella* or *Candolleomyces* fungus with herbs and spices is, Haitians know, rice at its best. So far, the first Black republic's foraging and cookery were the only thing keeping me from giving an entirely aspirational talk about wanting to help fight mycophobia among Black diasporic people, so paltry was the evidence that other Black folks scattered about the globe had taken any of the mother continent's mushroom love to their new lands.

Idly, I plugged the terms "junjo," "duppy umbrella," and "djon djon" into Google. When I did, something tantalizing happened. I was given a window to a past I did not know was possible before my mother offhandedly spoke the possibility into existence. Jamaicans (and those residing on the island in a colonial capacity) have been talking about the local funga for a long time. This fact, and others related to it, would point a way . . . not home exactly, but toward a tradition I didn't even know I was hungry to join.

All soil is made of ruined stars, but wherever I've landed, it's had a local flavor. My older brother perhaps understood this as he gave me a slice of mudpie to eat, with its sandy, South Floridian, salty crunch. *Here,* he might have been saying to five-year-old me with that uncharacteristically domestic gesture, *eat of the land and be naturalized to our adopted home.*

It became a habit of mine to eat dirt. I never ate very much, and was an otherwise healthy, if melancholic, child. I was angular, a starer (despite uncorrected myopia), and, among strangers, a half-whisperer in my waning Jamaican accent: a child destined to stay strange.

In that immigrant kid headspace with its compound imagination, inside which happiness *didn't* depend on the others my age who also just happened to reside in that corner of rural South Florida in the early '90s, I hosted a cast of characters who would not let me be lonely. These made-up people could surprise me—for example, asking to hear me read a verse from my blue leatherette pocket Bible, my best token of my grandmother—but unlike the children I really knew, they took my worthiness to play with them for granted. If I was spending more time with constructs of my own mind than with flesh-and-blood classmates, maybe the geophagy was a way to bring my seedling body down to earth and ease the stress of an overseas repotting.

Whatever the dirt-eating did for me then, I now know what beachy, sandy soil tastes like; I know what black, rich soil tastes like; I know what clay-red loam tastes like. I've sampled it all, and remember. According to William Bryant Logan in his book *Dirt,* that's a lost art—a farmer used to be able to chaw on a lump of soil and tell how it would serve a certain crop.[2] Perhaps my stepfather, who ran the farm I moved to when I came to the US, could have employed me as a soil taster if we'd raised vegetables rather than animals.

Though that particular brand of pica has mostly worn off for me, the fascination with dirt persists. I'm a sucker for the bizarro French manicure that comes from digging with one's bare hands, and I only cleanse my fingernails of that good, clean dirt so that I can enter polite society. When gardening, I'm easily distracted by how an earthworm's translucence lets you watch the soil move through them. Planting gets done only slowly. I will part with large sums of money for a candle that uses essential oils to mimic petrichor, the smell of rain-soaked earth. I want to be able to smell that newly wet dirt in the middle of drought. Nobody buries me in sand up to my neck at the beach anymore, and I miss it. That skinny, mudpie-eating weirdo grew into an adult with earthy predilections.

Given all this, perhaps it's odd that fungi took so long to find me. Though actually it's truer to say that it took a while for me to realize they'd found me; eating dirt means I might have been consuming several yards of fungal filaments with each mouthful.

Fungi, to be clear, are not metaphors, though some have used them as such. They are wholly themselves, nothing more and nothing less. But they are experts at knitting together those things that, on a cursory glance, appear discrete.

Like the eras in one's life. Not to be dramatic, but in the course of my own existence, there is the time before fungi and there is the present. So many aspects of "before" only make sense to me now through the magnifying lens given to me by the search for fungi. Before fungi and after fungi was not a mental border consciously erected—it's corny enough that I kinda wish I could resist it. But corny or not, it meaningfully describes how I used to show up in the world and how I show up in it now.

Let's pencil that border onto 2017 for the moment, a year when I was still right in the middle of, or at least thought I was in the middle of, one of those past lives people are always talking about. Trump was newly in office. For the US, this was like being diagnosed, as a teenager, with a terminal illness it turns out you'd been born with. So many around me were off-gassing misery. Their despair felt like it was missing the point because it mourned a kind of democracy, or people power, that we've never had. Secondhand wretchedness was thick in the air. And I was the most isolated I'd ever been.

It was becoming clearer by the day that the partner I lived with had a vision for his future that was incompatible with mine—that I was a weird, queer, cyclist peg attempting to fit into an impeccably landscaped, violently hetero, roads-have-no-shoulder hole. The career I wanted did not seem to want me, and my attempts to make moves to correct that felt like spinning in place. All around me my peers were making chosen and biological families and building activist networks and sharing their art. In doing so they modeled for me what I wanted, but there was always a voice in my head saying that they'd gotten there because they knew who they were, a privilege that was not available and could never be available to me. Those who have felt like the owner's manual for their three-score-and-ten was not included in the box will know what I mean. Maybe "who I am" was buried under a morass of frustrated expectations and goals for making art and loving on my own terms that felt delusional in an apocalypse. Unable to see clear to that fabled version of me, which was really a chimera of the achievements and expectations of others and therefore not truly inhabitable, I waded into an ocean of cheap cabernet to distance myself from my day-to-day. I kept living the lives of others—riding the train into their towns to walk their dogs and look after their homes, ghostwriting their books, falling in love with their partners-to-be, a means of fighting isolation that further isolates.

My camera roll from around that time starts off equivocal. There is a photo of a hen-of-the-woods mushroom from the city, taken, no doubt, to check with my Polish friend about whether I had what I thought I had. It's the only mushroom picture from the 2016–17 season, and it is very much a culinary-only affair: not taken in situ, no natural light, no loving angles, just the view from above it on the kitchen counter. The beginning of 2017 has so few pictures it's like I wasn't living at all by the day's standards, a result of my fake-Luddite keyboard phone's shitty camera, which did not reward bothering. Still, there are a few shots of the betta aquarium, one of Jim Dine's *Two Big Black Hearts* at the local sculpture park, one of my mother giving bombastic side-eye (a photo that I joked with her I would use, in her absence, to keep me on task), a few selfies I'd taken that show an impossibly youthful face smiling, plumped from crying, framed by the locs I'd just asymmetrically cut in a fit of pique. There are pictures of signs and banners from the March for Science on Boston Common, held in response to Trump's proposed cuts to all manner of science funding. "Be the solution, not the precipitate." A witch in head-to-toe black with identity obscured declaring, "My science is intersectional." There are several photos of passages from Alice Notley's epic poem of the underground, of orality, of citation, and of disruptions to "ongoingness," *The Descent of Alette*. In the first, the hero meets a dark-skinned goddess who makes and remakes the world from mud and clay. The poem begins:

> "One day, I awoke" "& found myself on" "a subway, endlessly"
> "I didn't know" "how I'd arrived there or" "who I was" "exactly"
> "But I knew the train" "knew riding it"[3]

Later, in autumn, that Polish friend of mine who had grown up hunting mushrooms and acted as my foraging mentor had a dinner party and sent her guests home with a peculiar favor—a paper shopping bag filled with as many hen-of-the-woods mushrooms as they cared to schlep away. Under her tutelage, I'd picked, cooked, and enjoyed the sculptural fruit bodies before. But it wasn't until that evening, during a wine-drenched moment when I stole away to really look at the dozens of large mushrooms she'd lined up on a bench near her back door, that I realized *Grifola frondosa* was more than inspiration to get creative in the kitchen. I picked one up. It was the heftiest I'd ever held, weighing at least fifteen pounds. I looked closely at the patterned, leaf-like caps that give the fruits their descriptive Latin name, flipped it over and put my nose right up against the white fertile surface with its many pores, like the inside of a bone. This chitinous mass was actually stunning; the way the silvers and creams shaded into the brown zones of the pileus mesmerized me. I thought about the way their flesh squeaked upon being harvested with a knife. A deep quaff brought me the vaguely dirty animal aroma that translated to such rich umami when they were sauteed in cast iron. I searched for and found some of the fungus's tenants—an orange-striped salamander coexisting with a jumping spider on the mushroom's interior terraces. I thought about how the oaks they skirted were usually girthy elders and how the fact that this particular fungus grew there could offer clues about the land from which it emerged. If my usual move at parties had always been to find the dog or cat with which to splay out on the floor, I had now outdone myself by finding the mushrooms at which to coo, with which to cuddle. That night, *G. frondosa* had come into focus as something more than food, had become an organism: the mycelium, that being doing humble but essential work and making beautiful art in its spare time, a dramatics of survival in multiple acts. The fungus was using its scent and its taste of earth and its beauty and its heft and its eerie polyphonous voice to speak questions directly into my spirit. And my answer in that moment, because it was high time I immersed myself in a nonhuman language to the extent that I could, was, "Yes, just let me get my hiking boots."

With winter looming, I joined every online fungal forum, watching with satisfaction as posts from my past life were replaced on my newsfeed by fungus identification requests in fora I'd lately joined. I bought every book. The queendom I'd begun acquainting myself with was endlessly fascinating, omnipresent, all-important . . . yet relatively little was known about it. On days when I didn't have dogs to walk, I was out on the trails near my home, determined to tread all of the 250-plus miles of them, the thousands of acres readily accessible from my front door. I was gone sometimes from sunup to sundown, my health app logging 30,000 steps between 8 a.m. and 7 p.m.

Under the canopy there was no phone signal. The mushrooms hid when you first walked by, the photogenic troops periscoping from pulpy, broken wood on the side of the log that came into view only on the return. Day after day, I got on my knees to take photos of their textures and colors. I chewed and spat their flesh, the residue of their identities on my tongue. Learning each new species helped terraform and fertilize what had been razed in my heart's landscape. I was learning that if losing one's self in other people could mean sudden enclosure, a shutting out, a hastily scrawled "no trespassing" sign, then losing oneself in a mysterious, capacious, and ever-expanding science might mean the opposite: no gate could shut to keep me away from the world fungi opened up.

Another inhabitant of the woods behind the nearby college was a black and white spiny-backed orb-weaver spider, who spun a new web every night. She hung upside down from her temporary home and defied my phone camera's focus, sharpening before the lens only when I placed my hand in the field so that my palm faced her underside. When I did this, a trick of the wind through the leaves brought her closer to my fingers and when I started, the wind laughed. There are two crystalline portraits of her out of 100 taken. For her slipperiness, I took to calling her Anansi, imagining that hers and the mycelial webs below my feet were mirror images of one another.

My camera roll got a lot less equivocal after that. Of the 75,735 photos in my library, roughly half are fungal. The old gag is that we childless mycophiles have fruit bodies where photos of our kids would otherwise appear. There are dates in my life that I can only navigate toward by remembering which mushroom I photographed that day or week, reverse engineering a memory by using hyphae as the building blocks.

Here was a different kind of love to feel, perhaps the closest I'll ever get to the agape variety that was supposed to animate me in church growing up. Meanwhile, the partner I lived with couldn't forgive me for disappearing in a cloud of spores. I don't blame him; not only had I stopped cowriting the fiction of my domesticability, but I was once again going feral, turning him to a mushroom widower, inchworms vibrant in my hair, psilocybin reminding me that the day is plastic. One day in the woods, I told Anansi that I was going home. In response, I saw him take down his old web by the light of the sinking sun and set to work building a new one. "Who I am" spoke to me clearly in that moment. It was time for dismantling. I made a New Year's exit from that past life, but in many ways, I'd been gone for ages.

Here is the thing I found online years later, on the floor of my mother's house, the words "junjo" and "djon djon" caroming around my head. In 1958, the journal *American Speech* published an article by Irish anthropologist Robert Wallace Thompson entitled "Mushrooms, Umbrellas, and Black Magic: A West Indian Linguistic Problem."[4] In it, Thompson resumes an earlier attempt to puzzle out the potentially African root of the words "junjo" and "djon djon." This obviously great mind who thought like me had written a note in the *Jamaica Historical Society Bulletin* four years prior, reasoning that the words denoting "mushroom" in Jamaica and Haiti were too similar to not come from a common source. He does some fantastic sleuthing across the Caribbean and, using plenty of primary sources, links "junjo" and similar words not only to fungus but to dirt on skin (Jamaica; per my mudpie eater), to a supernatural force protecting against evil or the potion that performs that function (British Guyana), and to a child's toy (Dominica). The article "Mushrooms . . ." was apparently written in response to German linguist Manfred Sandmann, who felt that "djon djon" must somehow come from "champignon"—Black mycophilia not being sufficient to carry an African source word meaning "mushroom" across the Middle Passage. Thompson, of Sandmann: "He claims that we must thank the whites for spreading junjo in Jamaica. It was they, particularly in the seventeenth and eighteenth centuries who ate champignons." Here I had to go to Sandmann's original to witness the audacity for myself. The other linguist is writing in Spanish for the *Nueva Revista de Filología Hispánica*. It wasn't just academic shade on Thompson's part—old boy really didn't believe that the Negroes could possibly have their own word for mushrooms, carefully remembered from the before-time and passed on. In order to prove his point, Sandmann quotes "el gran naturalista Hans Sloane" who, writing in 1707, suspected that mushrooms had been introduced to Jamaica from Europe, since he only found two species there when he went looking.[5] So why would they have brought a word to designate these organisms, all the way from Africa, when white people taught them everything they knew on the topic?

Reader, I hope I don't have to tell you that there were fungi in Jamaica for millennia before Sloane was a fishhook in his mother's eye. Here is what Sandmann would have had to do in order to know that there were more than two species on the island:

1. Care
2. Hunt down and consult either *The Civil and Natural History of Jamaica in Three Parts* (1756) or *Hortus Jamaicensis* (1812). In the former, with its *ſ*'s instead of *s*'s, the author even says of "the white tufted oblique mushroom" listed under *Lepiota*, that "this species is very common after every heavy rain, and grows generally on decaying trunks of . . . cotton-trees: it is the only sort that is *in use* here [emphasis mine], and when warmed and pounded, is sometimes boiled with beef in our soups."[6] Perhaps it's not fair to Sandmann to dislike him for being a product of his time, and not having the internet, but life's not fair.

Anyway, our good man Thompson raises these objections to Sandmann's position one by one. There in my mother's living room, I practically stopped breathing when he cited the fieldwork of Frederic G. Cassidy, a Jamaican lexicographer born in Kingston and raised in Ohio, who in 1952 spoke with Jamaicans about their associations with the word "junjo." Those Cassidy interviewed had things to say about mushroom edibility. "A maroon, a fisherman, and a former *busha* (estate overseer)" could be the start of a "walks into a bar" joke, but they're also the people Cassidy interviewed, and they made the following statements about various kinds of junjo: "can eat; grow on tree," "two that grow on tree; can eat," "white people call it mushroom," "can eat some," "bark-tree junjo can eat."[7]

Can eat. Can eat. Can eat. Can eat. Four times. The ones that "grow on tree." Shared with a lexicographer, who may or may not have asked about edibility.

This small passage—included by the author to read someone who had disagreed with him for filth—had me dreaming. I imagined Jamaicans deep in the bush with handwoven baskets, foraging the usual black mangoes, Otaheite apples, and ackee, but with the tantalizing addition of *mushrooms*. I imagined an epigenetic capillary from the heart of a bushcrafting Jamaican or West African ancestor of mine pumping excitement at the sight of a flush of edible fungi directly into my own heart. I even imagined that electrical charge traveling from some tropical *Psathyrella* mycelium under her feet all the way to the far-flung black trumpet mycelium under mine, across place, across time. Absurdly, I imagined her using the delicate horn of the fruit body to telephonically tell me, "We eat them, too."

Ultimately, Thompson decided more research was needed before he could be less equivocal in his hypothesis. He felt a good place to go looking for further evidence might be among the Gullah populations of the Southeastern United States, a thought I swear I had also had independently as I was reading (am I somehow psychically linked to this deceased Irish man?).[8] Three years after Thompson's "Mushrooms" article, the Jamaican lexicographer whose research Thompson quoted published a new note that amounted to "Why the self-doubt? You were onto something!" Cassidy cites a proverb that roughly translates to "If you don't have meat, *junjo* will do," recorded by Thomas Banbury in *Jamaican Superstitions; or the Obeah Book: A Complete Treatise of the Absurdities Believed in by the People of the Island* (!) at the turn of the twentieth century, and sums up the proverb's significance this way:

> Junjo [the word], then, was in print by 1895; but, proverbs being, as they are, a distillation of folk experience, the appearance of the word in a proverb is several generations older—it or some other synonymous word. The only other word now commonly known is *mushroom* (the "parasol" and "umbrella" terms are sporadic), and several informants have made the distinction that *mushroom* is the "white people" word, whereas *junjo* is the "black people" one.[9]

He even throws a few guesses into the ring as to the forebear of "junjo"—suggesting that an Africanist might look into "juju" and "jumbi." The more I read, the more my head spun. Before my mother's lesson in Jamaican fungal taxonomy, our history with this kingdom of life seemed desolate as a droughty July. Now here I was, with new search terms to exploit. What connects us to the past is sometimes delicate as spider silk. I went hunting for other sources.

In 2007, the literary and critical studies journal for African American and African diasporic studies *Callaloo* published a thirtieth-anniversary issue in which the article "Jamaican Versions of Callaloo" appears.[10] In it, author B. W. Higman delineates what an impoverished understanding Jamaicans have of the term "callaloo" itself, since we almost exclusively apply it either to the fibrous leafy green *Amaranthus viridis*, or the dish made with it, often including saltfish, tomato, onion, garlic, pepper, herbs, and spices. Elsewhere in the Caribbean, though, especially on eastern islands, it refers to a rich soup made with okra, greens, and varied proteins like crab, turkey, pig, or possum. In the article, Higman suggests that this difference is the reason Jamaicans don't use the term "callaloo" proverbially, to speak about Caribbeanness, *créolité*, and multiculturalism. Basically because our callaloo is, no offense, boring by comparison, we didn't make sayings about it. I'm all for the light roasting of Jamaicans. Our self-regard can handle it. But the part of this elucidation that most interested me was this bit of knowledge the author drops, taken from a book of Jamaican sayings and a dictionary of Jamaican English, by the by:

> Finally, Watson in 1991 recorded "If yu cawn get calaloo, tek junjo," claiming that because callaloo is cultivated, and junjo (edible fungus) wild and less desirable, the meaning is that in life we must sometimes take what is available or affordable (162). You can't always get what you want. Here callaloo gains a little status in the hierarchy of foods, though placed only above the despised fungal growths. Earlier versions had run "You po' fe meat, you nyam junjo," meaning "If you are badly off for meat, you'll eat mushrooms" (Cassidy and Le Page 254).[11]

There's Cassidy, again. Once I had recovered from the impertinent editorializing of "despised fungal growths," I took in the provided picture. Watson does indeed suggest that an apt interpretation of the former saying in the passage starts from the premise that "callaloo is a *cultivated* vegetable while junjo *grows wild* and is *altogether less desirable*" (emphasis mine).[12] It is possible to both languish and thrive in this explication of the saying, if you feel a particular type of hunger. "Cultivated" is of course better because it means you have control over your food. "Wild" is of course better, because it means unless they have absolute control of the "wilderness," no one can keep you from eating.

Today, broadly speaking, Jamaicans are a mycophobic lot. The late, great Gary Lincoff suggested that any place with a British colonial inheritance tends to fear the fungus because the British did.[13] French "properties" may have fared better in this regard—see Haiti's black rice, one of the jewels of its cuisine. Even in mycophilic Italy, however, the internationally prized porcini were once known as "poor man's meat." Mushrooms have been billed that way all around the world, and I recently read an article suggesting that in Sri Lanka, another former British colony, as well as in the southern Indian state of Tamil Nadu, there were populations known to be aware of the edible mushrooms that fruit during monsoon season. The article's author proposed that these picking practices may have come in and out of favor with food scarcity trends.[14]

On its face, then, my mother's assertion that there were only two "kinds" of fungus on the island served to ratify Lincoff's generalization. Loquacious peoples, a category Jamaicans arguably fall into, name and name and name again the things that spark excitement, and a paltry two names suggests the spark never quite caught. But we're talking about Xaymaca, here! The Taíno word means "land of wood and water"—the three nouns in that phrase are some of the main ingredients in the production of fungi. Jamaicans are a curious, canny, and resourceful lot—like poor folks, like survivors, everywhere. And we're famously cantankerous—just ask anyone who has gone to a Jamaican restaurant and expected the usual service industry obeisance. Expect it all you want as you put in your jerk chicken order—you'll be waiting in vain like Bob Marley was for that one girl's love.

On the one hand, there is colonial inheritance. On the other hand, there is the legacy of freedom. The history of that jerk chicken you're still waiting to be served with a smile is a good example. The jerking process was born of rebellion and fugitivity, as a collaboration between indigenous Xaymacans and escaped Black maroons.[15] When you cook your wild hog underground, after all, smoke will not snitch on you, betraying your whereabouts to those who seek to re-enslave you. That you can cut and run with the meat you've thoroughly spiced and preserved means you're always ready to move. It stands to reason, or I hope is at least a fair hypothesis, that those inhabiting a place like Cockpit Country (Maroon territory), who have been so good at guarding their secrets and sovereignty, might have also resisted the colonial inheritance of mycophobia. In fact, when Richard Currie, chief of the Accompong Maroons, recently took his oath of office, he told an interviewer that the climate on Maroon land is unique in that it sustains "crops . . . not frequently found elsewhere in the Caribbean, such as berries, strawberries, and mushrooms that grow wild among the networks of springs and streams."[16] Wild mushrooms as food crops? That sort of culinary development doesn't occur overnight. When I reached out to Chief Currie, dying to know which wild fungi grow so readily in the terra rossa in that famously remote countryside, he said simply that there was "a wide variety." All right, Chief. Keep your secrets.

We know who wrote the books on "natural history" and who had a stake in protecting their traditions as a means of survival, passing them down by word of mouth until such a time as young ears grew unreceptive, called from the bush to march in lockstep with progress. British mycophobia could easily have overwritten a fledgling Jamaican foraging practice. Even California's 200-plus years of mushroom gathering is fledgling, compared to elsewhere in the world, all those ancient traditions. And trust me when I say that erasure is an ongoing process. On a different page in this book, we will watch together as a living Brit tries to overwrite the history of magic mushrooms in Jamaica before our very eyes. But for reference in the here and now, here is the type of Englishman a Jamaican mushroom hunter of old might have had to contend with—please enjoy author and mushroom enthusiast William Delisle Hay writing facetiously in 1887 about what he calls the fungophobia of his countrymen:

> They are looked upon as vegetable vermin, only made to be destroyed. No eye can see their beauties; their office is unknown; their varieties are not regarded; they are hardly allowed a place among Nature's lawful children, but are considered something abnormal, worthless, and inexplicable. By precept and example children are taught from earliest infancy to despise, loathe, and avoid all kinds of "toadstools." The individual who desires to engage in the study of them must boldly face a good deal of scorn. He is laughed at for his strange taste among the better classes, and is actually regarded as a sort of idiot among the lower orders. No fad or hobby is esteemed so contemptible as that of the "fungus-hunter" or "toadstool-eater." This popular sentiment, which we may coin the word "Fungophobia" to express, is very curious. If it were human—that is, universal—one would be inclined to set it down as an instinct. But it is not human—it is merely British.[17]

That last sentence is a sermon. It's probably also false modesty, since that same author wrote a novel imagining a utopian year 2180 in which only white people exist, having "wrested" "state after state and territory after territory . . . from wilderness and barbarism," in the process exterminating everyone else on the planet. While Mr. Hay clearly has a flair for the dramatic, the kernel of truth within his satirical description of British attitudes toward mushrooms is well taken. And it stands to reason that British colonial disdain for any natural entity that defies control would be transmitted, like a disease, to the general populace in the places they ruled. But the pockets where that annihilation wasn't complete, where Black Jamaicans retained their trust in, reliance on, and stewardship of the land, are telling. "Tek junjo." "Can eat." "White people call it mushroom." I wonder if there's a connection between British transmission of their hatred to their colonial subjects of these natural wonders, this nature on steroids, and their abject terror of being poisoned by these abnormal, worthless, tricksy, and inexplicable beings. Poisoned by people they know wouldn't mind if they died. Could there be a link to the fresh memory of all the plantation owners poisoned to death by Obeah practitioners during Tacky's Rebellion? Hmm. Something to ask Anansi about later.

I called the presentation I began putting together with some of this information "Mushrooms in the African Diaspor(e)a." It's a rather idiosyncratic tour of Black mycophilia across space and time, libraries, others' mycological expeditions, and of course the internet. I undertook it to create my own lineage out loud and dare anyone to try and stop me, a lineage that other Black mycophiles with interrupted or untraceable histories might be able to see themselves in. It starts with mushroom-centric origin stories for Earth (Ghana's tale of how mushrooms first grew) and the heavens (Uganda's story of the Butiko clan's foundation), and includes a song sung to exalt the practice of mushroom gathering (by Aka foragers in what is now known as the Central African Republic), and generational transmission of mushroom identification from a blind grandmother to her granddaughter in Zimbabwe.[18]

I must admit that I luxuriated in my survey of Africa for perhaps longer than the talk's title suggested I would, wanting to spend a lifetime with the Butiko clan, whose name means "mushroom," the totem, or symbol, of their tribe. This clan's origin story involves the men going hunting and finding no meat, at which point they crouch to gather mushrooms instead.[19] *If you po' fi meat, nyam junjo.* This story is also a cosmology—the family's patriarch dies in the course of it, and his foraging basket is emptied across the surface of the moon, giving her those "craters"—really, it turns out, patches of mushrooms. There is a taboo against the Butiko clan's eating of the *Termitomyces microcarpus* fruits because they are considered their kin, but apparently a Baganda phrase encourages members of other clans, for whom butiko is not a totem, to harvest and eat them. *T. microcarpus* is a bucket-list species for me—the termites in whose mound the fungus grows actively farm it in a mutualistic relationship that benefits both fungus and termite young.

When I came across mention of the Butiko clan in my reading, I instantly felt an odd thrill, because I had been calling mushrooms my kin for years. The only other instances of this kind of frisson that I can remember came when one of my dearest friends and I first started talking and realized we had almost identical life stories, or when Facebook's facial recognition tried to tag the photo of another Pinto, a half-brother of mine whom I don't know at all, with my own name. *Oh, right,* I thought. *That* is *my face.* But without being able to say why, I lingered in Uganda with the Butiko clan. I lingered long after my presentation was given, then lingered again after the second time I delivered it to a mycological society, then again after the third. I've given this talk five times so far, and always repeat the Butiko origin story in its entirety despite wondering if I should truncate it when I have less time to speak. Instead, I've trimmed elsewhere. I wonder if I could tell the story from memory now.

Something was gnawing at me. The foraging basket's contents on the moon. The shape and clustering habit of *Termitomyces*. I kept coming back to my message exchange with Josephine

Nakakande, a Ugandan activist and mycophile, who told me that *T. microcarpus* is still very much a delicacy, served to a woman on her wedding night. I kept coming back to *Psathyrella coprinoceps,* its shape and clustering habit, how these djon djon have traditionally been served at weddings and other celebrations. *Termitomyces microcarpus* and *Psathyrella coprinoceps* are both palish, clustering mushrooms that grow in little troops, and while there is no termite mound associated with the latter, it's not hard to imagine the hunger that would send you out on a limb to pick the fruit body that looks so much like those back home. And when you didn't die, to tell your children these were safe.

And again, djon djon and junjo sound so much alike and look somewhat similar, too. If you cyan' get meat, tek junjo. Hmm.

Curled fetal on sphagnum moss, I know the earth wants me back. I can feel the gravity of her hunger, how she's fixing to take the meat off my bones like a cartoon cat takes the flesh off a fish. Clean like that, so that just my scaffolding, on loan to me for these short years, bleaches slow in leaf-dappled light. I sit on my calves, shins taking on moisture through my yoga pants. An entire universe within ten feet of where I'm planted. If it weren't for the search for fungi, chances are the only time my head would be at this level outdoors is when I crouch to relieve myself while camping. But the search for tiny pins and camouflaged fruits, the tricksters, has brought me low, on my knees, my cheek against fallen leaves. I'm reminded of a Nepali hiking partner, her basket full of colorful chicken-of-the-woods, genuflecting back at the trees as we exited the trail. During our hike, she told me that despite this town's housing prices, which tell us who the market believes has a right to live here, the Land knows we belong to it and so reveals her gifts.

Now, I watch the squirrels nearby who don't read me as a threat in my stillness, wonder whether they would like to gnaw the calcium from my rib, no longer mine, a file for their ever-growing teeth. The young blueberry leaves are apple Jolly Rancher green, vivid as the unearned smile of a baby. Lowbush blueberry makes up much of the underbrush here, and come July pastel purple bird shit will encrust the blue granite. The granite itself is tattooed in whorls of crustose paisley lichen. It looks as if it's cosplaying mold. An owl pellet has rendered some former vole or wood mouse into a cylinder of bone and fur—and inside, a shocking ribbon of paper towel. I do a double take but manage to tamp down the dread when I see that it's a six-spotted tiger beetle, and not an emerald ash borer, that has alighted next to my boot. A fire that must be several years old by now made partial charcoal of a segment of oak branch—the foliate roundness of newer lichen growth against the geometric crags of burnt wood is the stage for a tiny spider's jerky hydrostatic movement. I watch as they eat an even tinier spider. In a crevice near this arachnophagy, the thatched brown fiber of a vacant insect egg mass hangs loosely from the wood like a stormed barn door. Cervid musk hangs in the air and deer scat, dried and fungus-less to my naked eye, reminds me to check for ticks later, the predator I fear most after man here in the Northeast woods. The presence of deer reminds me to check, though the burrowing rodents whose homes I stepped over to get here are likelier vectors. Seedlings, long thin adolescent trees with shining bark, the standing dead of decades-old oaks, a birch growing softly into the trunk of a freshly dead ash in a morbid hug, the name-tagged 150-year-old red oak whom I know for a fact lives with slow white rot, down the trail a ways. A Mylar balloon glares in the sun, and if I can remember, I'll add it to my basket with the only other harvest: a spiked SunnyD seltzer can. This is just one day, one hour, but it contains every other time that it has ever been on this patch of land. It holds yet unperceived clues as to what will be.

It's May. Fungi are year-round, but for some, the mushroom season is just beginning. My group chat with the other two Boston-area Mushketeers (yes, that's what I call us, no, I won't apologize) is full of anxiety about whether the early onset of the season means our mid-month "morel weekend" trip to the western part of the state will yield only *Morchella* fruits too dried-out, rotted, or slug-eaten to eat. This will be the third annual trip—last year I was surprised by how emotional it made me to be surrounded by organisms not often seen in great numbers in eastern Massachusetts but spoken about in hushed tones in naturalist circles—*Fraxinus americana* and *Fraxinus pennsylvanica,* the white and green ash. It's odd, but while the coveted mushroom was certainly my initial reason to study and get to know these trees, a sense of watching history unfold has pushed that study further as I learn how the burrowing beetle, the emerald ash borer, myceliates its way from where it was first spotted in the US, in Detroit in 2002, outward over the country.[20] It arrived in Massachusetts in 2012, and the precious time we have left with the ash has rendered the morels it partners with very nearly beside the point.[21] North America has lost several of its beautiful and ancient tree species, to logging, chestnut blight, and Dutch elm disease (the latter two both fungal). Learning what our forests used to look like before these souls became endangered species makes me wistful for a woods I never knew.

Fungi have given me regionality and seasonality. They have given me land, wood, and water, the language to describe habitat. They have given me home in the way I used to know it as a child—that place where you essay to understand what Earth is made of, experimenting with its textures and boundaries, testing where you begin and the Land ends, or where the Land doesn't end but renders you continuous with it. I eat this holy dirt again in a poorly cleaned black trumpet (or when I go tumbling down a cliffside). I find unexpected plenty that comes home with me in a basket improvised from a sweatshirt, I give up the hunt without my quarry and must buy the mushrooms for the recipe. Fungi have given me back paradise, if it could be said I ever had it. Here's how: To see fruit in trees lets you know it's going to be all right. The immediate future is provided for. That's what going into the woods does now; I see the energy, sweetness, flavor, and spice growing from the ground. I know there are things I can do to keep a body from starving. Methods of survival that my ancestors knew that I could learn. Fannie Lou Hamer said, "When you've got 400 quarts of greens and gumbo soup canned for the winter, nobody can push you around or tell you what to say or do."[22] On the horizon, there's a return to these ways. A diversification of modes of nourishment, and medicine. You tend the crops and I'll reap where I did not sow and we'll can the bounty together. The study of mushrooms taught me to see plenty, abundance, and potential where I'd been taught there was none. It also taught me to not be frightened if there isn't plenty.

While preparing that first iteration of the "Diaspor(e)a" talk for a Midwestern mycological society, which I would deliver over Zoom at the height of the COVID lockdown, I corresponded with a Jamaican chef living in Mexico City who admitted feeling reluctant to post pictures of mushroom dishes on her catering service's promotional page. She feared other Jamaicans would heckle her. But all it took was living in a fungus-love capital of the world, and another yardie loudmouth declaring her own love publicly on social media, to make her reconsider.

"You gave me a boost," Theresa Meza wrote to me, "because in JA it's associated with duppy, and being exposed [to more of a foraging culture here in Mexico] I now use it." Despite that deathly association, she wrote, she put fungi on her catering menu. She sent photos of plates with huitlacoche nestled next to sweet plantain and curried portobello in coconut milk, among other combinations that made my mouth water. Theresa told me that a friend had sent her a news story about my guided mushroom walks, which bolstered her own burgeoning love of mushrooms. It made her question why she was so worried—were we not, as a people, known for culinary invention, for making whatever ingredients were available sing? When I asked whether she ever put any mushrooms in dishes on her regular rotation, beyond the catering, she replied, "I do but for my vegans." Here I felt duty bound to share the evidence that Jamaicans had a hidden culinary history with mushrooms as "meat replacement," specifically. Theresa was surprised and pleased. She had only ever known junjo in the sense of mold, as in "Junjo tek up di bread!" and so it was strange to see a more expansive history for it. But a crumb dropped in the woods, even one *weh junjo tek up*, can lead you back home. I like to imagine a new iteration of her menu with the heading: "For My Vegans: If you cyan' tek meat, tek junjo." But all over the Caribbean, there are traditions in need of new champions.

The comment section of any online news item, especially one from fourteen years ago, is not the kind of place one likes to linger if the general ambition is to live well. But linger is what I'm doing this morning, on an exchange between users Cerberus, Renard, and MangoSweet from a page on *Dominica News Online*'s 2011 photograph submission guidelines.[23] Instructions are given under a photo of a beautiful *Ganoderma* flush—its caps look like an indeterminate psychedelic sunset with zonations of crisp orange, yellow, crimson, and peach. But it's the comment section, robust as any community message board, that has my attention. Cerberus has responded to a comment from Renard, thanking him for sharing information about the pictured "lingzhi," disclosing the fungal happenings in their garden—how some of these weirdlings sprout from old trees and others directly from the soil. They want to know how to tell which are edible and if there's a field guide to the fungi of Dominica. As I read, I think, *There will be, someday soon, which my friend Melinda will author.*

Before Renard gets down to properly answering this question, he must disclose that he does not eat mushrooms. Not even in restaurants or canned soups! He goes on to do some serious generalizing about his country, asserting that Dominicans have never harvested mushrooms for the table. If they had, if these things were a viable food source on the island, "the old folk would have taught us to identify the good from the bad." He recounts a story from the '60s in which silly tourists collected the wrong mushrooms and became very ill after eating them, but he made sure to distance Dominicans from such foolishness.

But then MangoSweet comes in strong: "There is a saying in our local Kweyol that translates to: 'What you don't know is always older than you.'" With that, this stranger has held out her proverbial palm and I am ready to eat from it. She proceeds to school Renard, saying that one need only ask those in rural areas, who will say that they used to pick a fleshy white mushroom from decayed white cedar and toss it in the pot to make sancoche soup. And there was also a bright orange mushroom that grows on "Karapit trees in the rainforest"—they could also be eaten without dying. She says that Dominicans do know the wild mushrooms that they can eat and they know which ones to avoid, too, like the ones that grow on *kaka-bef* (!). I hope she said this last winkingly, with a little chuckle as she typed.

My friend Melinda Christian would recognize the value of the above exchange as an ethnomycological text. The New England mycophile's father hails from the East Caribbean nation, and she worries that, in the reasonable sprint to modernize and assimilate to the pace of the global economy, knowledge of the old ways will be buried, or sometimes drowned, in the scramble.

"It's heartbreaking to think of all that was lost," she told me, referring to botanical archives that were destroyed in Hurricane Maria (a storm that is five times more likely in the climate of today than that of 1950).[24] She'd been in touch with officials in Dominica in charge of these kinds of inheritances, writing letters and recording oral histories as part of an effort to conserve what knowledge she could from afar. She's not some starry-eyed romantic either—she knows that losing sight of where they've been as they look ahead tends not to work out well for any culture. I imagine this sort of accounting feels especially important for a location this awash in organismic richness. That you could lose without even knowing what's being lost would cause a special grief.

Years back, Melinda gave me a five-second tutorial on how to say "Dominica." It's *DAA-muh-NEE-kuh*, and one must briefly luxuriate in that third syllable to get it right, giving it love so that the hearer's mind doesn't travel to that other Caribbean place and neighbor to Haiti, before one is through naming it. It's a common mistake, a mistake I made before I knew better, one that has caused some on the island to desire a name change. In fact every time I try to Google the place, my autofill acts up. The Dominica that gave rise to half Melinda's heritage is called a hidden gem for its rainforests, volcanic activity, rare fauna, and variegated biomes—its overall stunningness. The moniker was given when he-who-shall-not-be-named did his 1492/1493 sail-about and was like, "I am seeing this landmass on a Sunday. That's all posterity needs. Boom. Perfection," as if that makes sense, which it doesn't.

The Kalinago people, who had been displacing the Taíno there for a couple of centuries at the moment of European contact, call it *Wai'tu kubuli*, meaning "tall is her body," for its tendency to go straight up from its coastline, steep as the day it was born, which was not that long ago in geological time.[25] Melinda herself is six feet tall and regularly climbs New Hampshire's 4,000-foot peaks, sometimes with a small child strapped to her back. The Kalinago, known as the fierce warriors who repulsed Spanish advances with a "poison arrow curtain," leaving the English and French to duke it out and stake their claim, number in the thousands on Dominica, earning the distinction of largest indigenous population in the Caribbean.[26]

Melinda's father let her interview him about the fishing, hunting, and foraging they used to do on the island to survive. He has not always been so forthcoming, and she theorizes that the shame of what-one-had-to-do-to-keep-body-and-soul-together was a painfully vulnerable place from which to share. But when she persisted in her study of mycology (which commenced when she was in high school, then lay dormant till she found the online community where we virtually met), he relented, telling her about the honeycomb fungus and the cone-shaped mushroom that he and his brother, at ages five and seven, would forage, part of a wider food-gathering campaign that included fishing and growing vegetables. Feeding a family of eleven is no small task, and of course foraging whatever one could would help ensure that bellies were as full as they could be, with meals as nutritious as possible. As part of her campaign to document Dominican funga, Melinda founded the Facebook group "Waitukubuli Fungi (Mushrooms of the Commonwealth of Dominica)" in hopes of showcasing (and archiving) "the glorious biodiversity of 'The Nature Island.'"

Given her description of the beauty of the place in an email, I hope she'll accept me as a field assistant on one of those trips:

> It is magical, especially in the summertime—storms launch the smell from the bay leaf huts into the air over the bright "cirique" (tsee-week) crabs scuttling across the roads. Rainbows over Victoria Falls are commonplace and clean water is not a precious commodity. Seeing the ocean and the Caribbean Sea meet at Scott's Head with clouds rolling over the angry former is one of my most awe-inspiring memories. It's as if some steady hand left an invisible partition between placid turquoise paradise and Atlantic wrath. Like when painters use masking tape on a canvas.

Mycologist Lauren Ré gave a talk called "Mycology in the Tropics: Fungi of Guyana & Trinidad,"and I had to (virtually) be there.[27] In it, she describes an expedition to the very edge of South America, and to the larger of the islands that make up the nation of Trinidad and Tobago in the Lesser Antilles. After making collections in locations like El Cerro Del Aripo mountain peak and the Aripo Savanna, she and her team deposited 134 new specimens at the University of the West Indies, St. Augustine, campus in Trinidad. It was the first such deposit that the specimen library had seen since 1947. While the plant material in the herbarium there was extensive—tables of boxes upon boxes holding large folders, and heaps of cabinets—the fungarium within this collection fit in a shoebox. Considering fungi were not formally separated from plants into their own kingdom before 1969, this is perhaps not a surprise. From appearances, the last person to contribute to the library had submitted *Cyathus*, or bird's nest fungus—documented in January of 1947 and misfiled as a lichen. The neglect! According to Lauren, a former curator had not long ago declined to allow a mushroom-enthusiastic researcher to work with fungi for the collection. When I asked why the heck not, Lauren wondered whether the curator was wholly qualified for the job. Possible nepotism, definite mycophobia.

The fungi she and her team saw in Trinidad included the usual cap-and-stem jobs, the sand-loving waxy caps, the hairy polypores eaten by the Amerindians, and spectacular insect-infecting organisms sending up their fungal fireworks. Among the more mind-blowing phenomena they documented was a species in Marasmiaceae that doesn't make mushrooms, instead staying anamorphic and forming a web in the tree canopy that catches leaves like Anansi. The fungus then digests until just the leaf's central vein is left. It wraps these veins in noded matter that birds will collect for their nests, providing a nice antimicrobial place for their young to hatch.

Even more exciting for me was her observation of leaf-cutter ants tending to their fungal gardens as a way to feed their larvae. The ant brings the pieces of leaf back to the nest, which the fungus will chemically vet, sending out stress hormones if it can't eat the proffered plant matter (a preference the colony will then learn, never to make that mistake again). The fungus (which doesn't produce mushrooms because that's a lot of work and it has coevolved for 30 million years with these ambulatory organisms that just *love* to work, so why would it) will then predigest the plant matter for the ants' young. Learning this during Lauren's excellent presentation made me wonder whether such fungal gardening happens among Jamaican ants. Later, I was thrilled to find that it does, that in fact there is a species of "lesser fungus growing ant" first described in Jamaica in 1891 as *Trachymyrmex jamaicensis* (later reclassified as *Mycetomoellerius jamaicensis*), and they apparently feed their fungi leaves from the trees that produce one of my favorite fruit: guinep, or Spanish lime.[28] This is not the first time, and I suspect it won't be the last, that I've had to hand it to fungi for their impeccable taste.

To return to my own symbiotic relationship with mushrooms: I experienced the autumn of 2019 in the Northeastern US as a period of drought, though the official record shows none was declared. It could be that the explosive fungal activity of the prior autumn season spoiled me and the other mushroom hunters of my acquaintance, so that we all felt the crackling dryness of the land under our feet was abnormal—certainly, we were vocal about the lack of mushrooms online. Whatever accounted for that feeling of scarcity, I was relieved to arrive at my grandmother's home up a hill above Kingston that December: my grandmother's home, with its fish-tea air and the fog that hangs over the lush jungle of the backyard gully at sunrise. When I rose from a nap on the morning of my arrival, I saw *Auricularia* (wood ear mushrooms) growing from the ackee tree in her backyard. I chattered excitedly with my uncle about wanting to see what else was growing back there, and one morning he pointed out some fungi that he'd seen come up before.

"You know is what?" Uncle asked.

"Hmm."

I did not, so I took it to the internet. None of my guidebooks would have much to say about what we'd found—there is no fungal guidebook for Jamaica. In one online identification forum, I posted pictures of the specimens from several angles, along with the caption "warty, white spored, pleasantly floral smelling mushrooms growing from dead mango tree in St. Andrew, Jamaica." This online network is one to which I've entrusted my health dozens of times, and I knew they would never let me down (unless there was a disagreement among the old-school experts and wunderkinds about what I had). It didn't take long before an OG who makes his home in Ecuador replied, "*Oudemansiella canarii* or similar . . . the natives have eaten those here—it is not bad." Though they don't resemble one another too closely, a member of that genus back home in Boston is one of my favorites to find because of its deep rooting habit (try to get the entire stipe from out of the ground in one piece!), and an underrated edible, so I knew what had to be done. Shortly after confirming this ID using Mushroom Observer, another online resource, I sprinkled salt and pepper over the caps and crisped them in coconut oil. They retained some of their floral aroma, but the dominant flavor was funky, earthy umami. Sadly, my uncle declined to try his find, but I had better luck with my then-ninety-seven-year-old grandmother. She took the tiniest piece and smiled.

"Is it nice?" I asked her, laughing. She looked at me indulgently and gave the faintest nod. "If you could be sure it was these, would you consider eating them again if they grew back?" She shook her head no. It was a small gesture, but a decisive one.

I was once asked by journalist Ari Gray—who's also my occasional hiking buddy—something along the lines of what it means to me to be Black in the woods. I don't remember exactly what I said, but it was probably something about the relief I felt in shedding the racialized shell like a cicada's when I hit the woods—how I'm just kinda free until another human perceives me. But that was years ago. The way I think about my Blackness has changed. Mine is a Bredda-Nansi Blackness, a needing-to-know-and-to-collude-with-or-(rarely)-conspire-against-every-single-animal-in-the-village Blackness. I come from a tradition that incorporates every player, storifies every relationship, whose tales are each one of them a woodland vignette. I meet Bredda Nansi, or Aunt Nansi—him, her, them—whenever I go to the woods. When I go looking for mushrooms, I find everything else, including what fruits authentically from within me when nobody is watching.

Mycophilia helped me unlearn a tendency to vivisect and standardize that informed not only my idea of my place within (or outside) the ecosystem, but also my conception of how to inhabit my own Blackness. Before mushrooms, Blackness used to be so much less expansive because I allowed it to be defined for me by others, like cops, amidst concrete. The tools I was using to view my Blackness were of European manufacture, bequeathed by white supremacist schooling that so many of us must work to counteract. That race was socially constructed in these circumstances, but nonetheless informs the cultures one comes up in and the traditions one is pressured or allowed to claim or refuse, of course complicates matters. Amid these tensions, my Jamaicanness used to be distinct from my Caribbeanness somehow, which was distinct from my North Americanness, my whole self further fragmented by being twice diasporic.

Some will disagree that I am anything but United Statesian, given how young I was when I migrated here; others might suggest the extent to which I represent my Jamaicanness or even West Africanness to outgroups is the true measure of who I am. But the pursuit of fungi has me reconstructing my conception of the "self" from scratch—with pantry ingredients that spring back to life with rehydration, along with novel staples. What it means to be Black has taken a surprisingly central place in my rebuilding, surprising because it wouldn't have occurred to me to lug into the woods the socially constructed version of myself. But my Jamaicanness is made of fresh-picked food; of making the best of the ingredients available; of gloating with loved ones about how good it is despite its humble ingredients; of knowing how to work the land you live with; of a reciprocal relationship with that land. It's church on Sunday, and whereas my cathedral is now usually a canopy of trees, it's as strong as the community that we're building there, which has been growing hunt by hunt, genuflection by genuflection. It's in all things pointing back home, from the chicken-of-the-woods patties, to hen-of-the-woods jerked jerky, to "saltfish" made from fishy salt-cured *Hygrophorus flavodiscus*. It might be curried venison, jerked wild turkey, or brown stew squirrel. Sorrel made with sumac and wild ginger. Solomon Gundy from fresh fish caught off the Massachusetts coast. Drumskins from amadou. The list goes on. A child of the diaspora, I recognize that every "new" thing is made from ingredients gathered before. I insist that my identity did not spring from a colonizer's head.

The study of mushrooms has given me back some of my ghosts. If most ghosts are gauzy apparitions, mine are slightly more vivid now, somehow. I can see the connections straining to make themselves to me. Some have said that whatever happened before colonizers came is of no comfort, and that feels right to me—with the proviso that whatever happened before is a wind I will happily feel at my back, once the ghosts there are speaking a language I can understand.

Histories of migration, questions of who brought what where, did not come fully alive for me until I found the sticky web of ancient, ever-present mycelium. They can chop down family trees, even hang us from them, but the mycelium keeps spreading under their feet.

Something still gnawing at me recently, I found myself thinking about the Butiko clan again, those mushroom kin. As distant laughter sounded through my open window, I mixed search terms about the Butiko like it was a word game I might eventually win. Sure enough, I found a page I'd never seen before—a blog post on beingafrican.com that said this under the heading "Butiko Clan (mushroom)": "The Akabbito (secondary totem) of this clan is Namulondo the clan head is Ggunju."[29]

The clan head is what now? Please, abeg beingafrican.com, come again? The Jamaican chef in Mexico, Theresa, had pronounced it "junju," and during my internet peregrinations, I'd see several spellings ending in "u." What if those in what is now called Uganda exported elements of their culture to Africans who were forced to migrate? What if those who were forced to migrate referred to mushrooms, slangishly, as the *Ggunju,* head of that one clan, the ones who can't eat these trooping beige-ish mushrooms, but who taught us how? What if those forced to migrate had softened *Ggunju* to *junju,* sometimes heard as *junjo*? What if, what if, what if, what if. The laughter sounds again, and this time it takes me a minute to realize it's my own.

IN SEARCH OF

Tuber melanosporum

(THE BLACK WINTER TRUFFLE)

On what lay buried,
the atmosphere of an alien
planet, that scent!, what is
holy, reclamations, labors of
love, the microscopic world,
letting the land dream

The most learned men have been questioned as to the nature of this tuber, and after two thousand years of argument and discussion their answer is the same as it was on the first day: we do not know. The truffles themselves have been interrogated, and have answered simply: "Taste me and you will be worshipping God."

—Alexandre Dumas

Is that all there is? I thought, the first time I had a proper sampling of a truffle. It's mostly earth and funk with nothing at its center. Very expensive, empty, earthy funk. A metric ton of what looked like a lump of coal had just been shaved over tagliatelle I ordered at a hip Italian restaurant in Cambridge. My friend was a cook there and kept sending freebies to my table, so it was kind of bratty of me to notice the expense, but skepticism is the natural state of those of my class, especially when we're assessing one of the world's most expensive foods. This "truffle" tasted nothing like the coat-your-mouth, oily essences that had been obligatory on menus in 2010 as a dressing for French fries (along with rosemary and parmesan), and I was disappointed. My naive palate had been expecting that chemical assault to the *n*th degree. I wouldn't learn till later that truffle oil is more a clumsy ode to reality than any kind of faithful distillation of it.

Talking to truffle fanatic friends later, I realized I had been barking up the wrong *Quercus ilex** at the Italian restaurant. It's not the flavor of a truffle that grabs you by the collar and makes you pay a dollar—it's the aroma. What I should have done at that bonanza of lovingly curated food was bow my head over my plate and take a deep draught, barely waiting for the server to finish shaving smooth, paper-thin wafers of marbled gray and white gleba with the thinnest halo of black, well-melanated peridium onto the dish. Really, one shouldn't need to put one's nose up against the plate to experience a truffle as gastronomically intended, but many of

* A species of oak with which truffles associate.

the imported specimens found in US restaurants are too many days out of the earth, so the chemical stew that makes them great has long faded in pungency.

Years later, finally wise to the truffle's particular charms, I made good and fell in lust with these earthy lumps, these fruits "consecrated to Aphrodite," as Dumas said.[1] My friend Tyler, a fine food purveyor and owner of the Mushroom Shop in Somerville, supplied me, at cost, one small black truffle in a plastic jar on a humble paper towel. Because the fruit was regularly exhaling its unique complement of volatile organic compounds with each moist breath (yes, truffles, like most fruit bodies, respire; fight me), I had to change the paper towel regularly. It was like shucking a gym shirt that had been sweated through, only you don't want to launder the paper towel. You want to keep it, and sleep with it next to your head on the pillow.

In that little jar was the aroma that had enslaved kings.[2] From the moment Tyler brought it to my place and I took that first whiff as he watched, it was over. I could not stop unscrewing the lid to sniff it. I might have sniffed it fifty-seven times in the first hour of our life together. A conservative estimate. That scent! What *was* that?!

If this were a few years ago, before a recent study, and I were the kind of female boar who had read previous studies, I would have said "that" was the male boar that I wished would please impregnate me yesterday. I would have said this because one of the compounds released by these fungal agents of chaos is similar to androstenol, a pheromone that emanates from both pigs and human men, among other animals. While there's no denying pigs were the original truffle detectors for humans (our living, snuffling olfactory prosthetics who sometimes got away with the loot), it's not necessarily because truffles give off aromas of a boar in rut. Go figure. It turns out both boars and dogs may be more attracted to dimethyl sulfide than androstenol, which sources say is reminiscent of cabbage, fish, and canned corn, of all things.[3]

Whatever the mechanisms of the attraction, it's clear that truffles are compelling to a diversity of beings, from the truffle flies who lay their eggs in overripe specimens, to the humans who spend big bucks chasing the mood-altering effects of the truffle's "bliss chemical," anandamide, to the bacteria that live with them as symbionts, to the squirrels, deer, bears, pig-boar hybrids, and more who root around for them. Black winter truffles have evolved to manipulate us by emitting 3-ethyl-5-methylphenol, 5-methyl-2-propylphenol, β-phenylethanol, and 3-ethylphenol, among other gasses.[4] In other words, sex, dirt, garlic, petroleum, ozone, olive brine—the list of words used to grasp at the experience goes on. A housemate said my truffle smelled like the atmosphere of an alien planet. Another absolutely required an essential oil from exactly that smell in order to make a candle (her pandemic hobby). My partner, Walter, said it made him feel drunk. Everyone came back for several more deep breaths over the little jar, like they needed to confront something that had been nagging at them. To my nose, the black nugget gave garlic, motor oil, soy sauce, wine, and acetone, all beneath an overtone I couldn't name. It reminded me of going to the gas station as a kid. Petrol smells less strong now because it's processed differently, but back in the day, its pungency would have me breathing deep till I was lightheaded, looking through the back windshield as my mom pulled out of the filling station, whispering "till next time." Over the next several days, I offered up a whiff to anyone who would humor me: most closed their eyes (you have to)—everyone paused thoughtfully as they sniffed, smiled the smile of one recalling a powerful memory that is nonetheless reluctant to fully form, then puzzled over what exactly *that* was. It was only when I urged them on that my subjects would fix words to the olfactory experience.

The next morning, ready to reencounter another of the truffle's aspects, I shaved some of mine onto steaming scrambled eggs. The heat from the protein had the effect of sending *Tuber* fumes into my face (enough to make me groan), and I took a great, big, greedy bite. And reader, what did I taste? What played on the ol' buds like it had always known them well?

Nothing, and a bunch of it, at that. It was as if the fruit body's essence had evaporated upon contact with my tongue, like I'd been had by a practical joker. Or been touched by a spell, daring me to explain how a scent so intense could be so imperceptible in flavor. In just about every other encounter with food, flavor is generally determined by scent. But my partner could taste the absence, too, like an edible riddle. Deus absconditus. Or a lover who will whisper you to the edge but who simply can't touch you, not where you want them to, anyway.

An industry insider I spoke with said that this was *the* open secret among purveyors and their clientele—the truffle is an experience, not a food. You're buying an idea, a brief essence, a marker of luxury, the notion that someone was out in the woods to hunt this for you with a dog. It's sort of amazing how much cultural cargo these nuggets of empty, earthy funk can carry.

Alexandre Dumas, of having-written-*The-Three-Musketeers* fame, also loved to write about fine cuisine. He called the truffle "the gastronomic holy of holies."[5] I get it: there's something sacred about the ability to be stumped in this day and age, to live, for one deep breath, in a question. Truffles trade in the ineffable. Their currency is scent. Unlike other sensory information, "smell is hardwired into the limbic system," writes Rowan Jacobsen in *Truffle Hound*, the obsessed author attempting to give a neurological explanation for why it's so difficult to process the hold truffles have on us: "Scents bypass the higher brain, instantly imprinting on emotion and memory without interpretation."[6] This is one way to say their aroma can inspire mindless longing, that they practice a form of olfactory hypnosis. Hypnosis has been used to remember. To suspend its subject in a state between presence and memory. Perhaps some of the truffle's power comes from how, through scent, it conjures the hint of a taste on the tip of the tongue, like a delicious dream whose outlines we only vaguely remember on waking. A dream which nonetheless has the profoundest effect on our bodies.

Before we became such a complex profusion of cells, we were driven by chemical impulse, which a truffle knows better than anyone. That's why they fetch the prices they do. Why animals will harm themselves to unearth a fruit. Why truffle hunters are poisoning one another's dogs.[7] Why suspected truffle thieves in French orchards get shot.[8] Why I brought that truffle everywhere I went for days, even driving with it down the entire Eastern Seaboard, like a talisman, like a security blanket, like a holy, holy tribute to the primordial cells in me that have operated under impulse power toward their chemical quarry since near the beginning of time. Why I asked a stranger from the internet if I could come visit her truffle farm in the Virginia Piedmont.

With her bell-clear voice and a tone about as sober as the one she uses when discussing scientific matters, microscopist-turned-farmer Jasmine Richardson sings Peggy Lee's '69 anthem of the unimpressed: "Is That All There Is?" She wonders whether that's all there is to a fire? All there is to the circus? All there is to love? If so, let's pop a bottle, basically, and, well, turn up. It's the third night of my visit to her nascent truffle orchard in south-central Virginia, and we are not totally sober, singing YouTube karaoke despite her crappy internet connection, our beer and wine emptying quickly. It feels good to croon together after a long day of weeding, crouching in her tree plantation and fretting together over the dozens of juvenile trees just outside the front door. I've suggested karaoke to gauge her gameness and continue to accelerate the process of learning about her, as I believe you can tell a lot about a person based on their song picks. "Is That All There Is?" is not what I was expecting—Jasmine's internet persona at the time was all smiles—but the surprise is intriguing.

One thing that *has* impressed Jasmine is the so-called Périgord or black winter truffle—*Tuber melanosporum*. She has pledged allegiance to *T. m.*'s mysterious allure, staking her future on it. The self-described former army brat, who lived in the UK, Kyoto, North Carolina, Berlin, and more over the course of her upbringing, always considered physical movement as a measure of progress in life. That is, before she began putting down literal roots in establishing her orchard.

But the gag is that this enterprise is not a sure thing. That's despite the roughly $50,000 in investment capital that she and her family will have spent—not including the purchase of the foreclosed property where she lives and where operations are based. Orchards of this kind are not guaranteed to take, even with all the resources that must be sunk into the soil to set the stage, the terraforming required in a mid-Atlantic context to gestate alien mycelium from the Mediterranean. The difficulty and capital-intensiveness of the proposition is one reason so many North Americans are surprised when I tell them Jasmine's orchard exists. In the popular imagination, truffles are associated with night hunts led by pigs or water dogs in France and Italy. It's hard to imagine their enclosure—not least because if truffles were easy to enclose, they would be much less expensive. But truffle farms have existed in one form or another for a couple hundred years. It's just that the formulae for their success aren't exactly common knowledge—by design. Just like a hunter in Italy's Piedmont guards the secret of his trees with a gun, so too do Europeans, Australians, New Zealanders, and Californians who are making a successful go of truffle farming hide their methods.

It helps to have a mentor from Spain. But Jasmine is also tenacious. As her song pick suggests, she's profoundly bored by ordinary life. A Black descendant of sharecroppers and former student of global and sociocultural subjects, she divides her adulthood into phases. In the current phase, at the edge of rurality, she wonders if and how her party girl era might bleed into the prevailing quiet.

Jasmine is petite, striking, high of cheekbone and broad of smile, and could easily be described as "likkle but tallawah," if a Jamaican were doing the describing. From my vantage, while she may feel farm-bound, she also resides in the speculative space of the question she is asking of the future. I had been fascinated from afar by what I could discern of her from social media, how she might be waiting seven years or more from planting to first fruits. I was taken by the endeavor's slowness and her nerdiness, how it all appeared to be an experiment, a public bet on herself, a labor of love. I wanted to know how she arrived at the middle of her own story, what she saw through her microscope's objective, what her days looked like. If a friendship formed as a result of my crashing at her place, so much the better.

When I finally worked up the pluck to ask if I could visit the farm, she said yes but warned there was weeding to be done along with a host of other tasks. Weeding and whatever else might come up seemed to me like a fine trade for getting to witness the scientific bottling of magic. So on my first full day in Virginia, Jasmine, my partner, and I weed by hand under a mellow early fall sun, grasshoppers arcing into the air around us as our hands disturb their perches. Sometimes, our hori-horis connect with golf balls mostly buried in the earth, strange inversions of the hoped-for crop: *T. m.* skins erupt with tens of small, blunted pyramidal structures, where a golf ball is pitted. Jasmine explains that the former owner of the property used the wide lawn for a driving range. We also find troubling evidence of some variety of burrowing animal that will have to be trapped or dispatched—lest it, like any red-blooded mammal given the opportunity, treat itself and its progeny to an expensive truffled smorgasbord. I lower my face to a few inches above the ground and sniff. Mostly vegetal, somewhat marine. As I watch the mica sparkling on my skin where I've been covered in soil, I'm struck by the ways the earth here holds long, long, human and more-than-human memory in the same breath as our geologically young bets on the future. What will *Tuber melanosporum* become in this soil?

Shortly before Jasmine was bitten by the mycobug, she was living in California with a long-term partner. Their breakup sent her into a crisis. I can well imagine how a party girl's instincts might kick in at the moment (they certainly did after the dissolution of my own long-term relationship), how one might easily avoid staying still long enough in one's own mind to really see the structural damage left by a dislodging. When Jasmine's father drops by the farm later for a visit (in the slickest black Corvette I have ever seen), he intimates that the split sent her spinning. What drew her back from the brink was the prospect of the grand *Tuber melanosporum* experiment, along with one of the many varieties of psychedelic-assisted clinical therapy. I'm reminded of my own split and how I took to the woods on small doses of psilocybin. Jasmine tells me she listens to "Is That All There Is?" to prepare for the administration of her treatments. When I hear this, I realize that Jasmine is the perfect answer to every nineteenth-century novel with a brilliant brooding heroine at its center, dreaming of freedom.

Yet before I learned of Jasmine's ambition, I was uninterested in fungal cultivation. For me the search for mushrooms had always been about following my nose—the pursuit of surprise and novelty. A tolerance for uncertainty and the ability to modulate one's expectations were flexibilities I felt I needed to learn, and poking around in the duff looking for food can supply both in droves. As in: *Well, we haven't found matsutake but we did get to watch a snapping turtle eat a frog and we collected wintergreen to make a minty infusion, so the outing was a success*. This feels especially useful in a place and time when, for a fee, one can get anything delivered to their door in a matter of hours. Like some landed aristocrat of old. But the *T. m.* orchard also has its flirtation with the unknown, and risk, and uncertainty. Plus, these organisms are different. They're a dream, a hypnotic trigger object, a haunting, an essence without body. Even if you succeed in growing them, can you really tame the holy?

Jasmine keeps some of her perfume collection in the guest room, and that night I test the scent of each one. The result is a muddled miasma I hope will dissipate quickly. When I ask her about the bottles, she says that she's been collecting fragrances since she was in high school, a hobby she no longer engages in. Of course, she's now attempting to grow nature's consummate perfumier, as truffles have mastered the manufacture of aromas that become obsession. In addition to their staggering number of volatile organic compounds, they also tend the bacteria within their fruiting bodies to make a unique, fermented product many times more complex than the aromas of the finest of fine wines. Fascinatingly, it seems there is a correlation between the truffle "bliss compound" of anandamide and a black truffle's blackness. In a study from 2015, researchers proposed that the anandamides in a truffle control melanin synthesis in a process that both aids the truffle's maturation (during which its fertile surface darkens in color) and signals to truffle-hungry mammals that they should seek them out for the fruiting body's intoxicating effects.[9] Does this mean that the blacker the truffle the sweeter the fruit?

The scent we identify as "mushroomy" is a product of mushroom alcohol, or 1-Octen-3-ol. This compound, produced by the bruising and oxidation of fungi, and present in some plants as well, gives rise to the deep lustiness underlying most recognizably fungal aromas. It's a component of damp basement, wet earth, human breath, and sweat. It attracts mosquitoes, which may well explain why some of my most productive foraging spots cannot be approached without paying the itchy welt tax.

Mushrooms choose from an extensive scent library and trade on combinations that would only occur to the most ingenious alchemists. We smell something because it is sufficiently chemically reactive, it will bond with fat, and its molecules are duly small. Scents deliver striking pungency if several olfactory receptors are tweaked at once in novel or nostalgic ways. Because we have so few words to describe scent, but as humans we can detect 1 trillion odors (while about 1,000 live actively within our recollection), to think with smell means sometimes resorting to the language of the synesthesiac.[10] Jasmine describes the truffle's bouquet as *cozy*.

One of my canned lines when leading a foray is "mushroom hunting is a sensual experience." Sometimes this draws titters or smirks—folks aren't used to linking kingdom fungi with sensory pleasure outside of eating well-prepared mushrooms. But scent, already established as perhaps our most primitive sense, is also one of the most useful tools of mushroom identification we have. *Mycetinis scorodonius* (vampire's bane if you're nasty) smells more like garlic than garlic does. You better really love the scent of maple syrup if you're going to dry candy cap mushrooms on your dashboard—your car will smell like a pancake breakfast in New Hampshire in April till it retires to the scrapyard. The ashtray bolete smells exactly as the name suggests, and *Agaricus* species that go under the name "the prince," along with some of the fetid species of *Russula,* pack enough of an almondine kick to remind someone of their great aunt who loved amaretto and marzipan candies. There's the cucumber rind of dryad's saddle, the bleachiness of certain powder-veiled amanitas, the anise that shades into fishiness of oyster mushrooms, the cinnamon plus aged cheese scent of matsutake. As they mature, hen-of-the-woods smells like Fido went for a swim and now wants to sit in your lap—I know one forager who prefers them aged to this point. Smooth chanterelles in my region smell of apricot. Some say the green goopy gleba of stinkhorn mushrooms always smells of rotting flesh, but I know I'm not the only freak who sort of likes the way the *Phallus ravenelii* I left in my pack with my leather jacket that one time had the jacket smelling for several days. There were other, more pleasant scents underlying the "corpse" that made things interesting. Who was it that said there is no exquisite beauty without strangeness in proportion? Was Poe talking about a stinkhorn?

After another day of work, I make a giant cast-iron pan of shepherd's pie, featuring bison and hen-of-the-woods mushrooms I've brought down from Boston. Two of the men Jasmine regularly employs to bush hog and do other assorted tasks around the farm are invited for dinner, but only one of them, David (a taciturn young man of Chinese descent with a deep country accent whom Jasmine found through the online Virginia mushroom-hunting network), will eat a dish that features fungus. Mason doesn't eat mushrooms or bison but happily munches on the buttery mashed potatoes that top the dish. When I ask whether he'll relax his no-mushroom rule to sample the fruits of his and Jasmine's labor in the future, he smiles coyly.

Later that evening, Mason drops a bombshell. We're discussing the former resident of the Richardson homestead: how he built the large home, his dream house, complete with putting green; how he lost it all during the 2008 financial crisis; and how he slowly drank himself to death, with the basement bar as a haunted relic. Mason offhandedly mentions local lore that the family had some ties to the KKK. I can't tell whether this is the first time Jasmine is hearing about this speculation. She makes a joke about having a crest made that will show her stomping on a Klan hood, and I think again about what it would mean for her to succeed in raising truffles in this particular soil.

It's less than common knowledge that *Tuber* devotee Alexandre Dumas's grandmother was an enslaved Black woman, that his father was a famous general. His haters loved to try to level his Blackness against him. Balzac once visited his publisher's office with big "Bitch Better Have My Money" energy, and the publisher said that his pay had been equivalent to Dumas's for the same amount of work. Balzac replied, "From the moment you compare me with that Negro I have the honor of wishing you good evening."[11] Eugène de Mirecourt (who wrote, but whom nobody has ever heard of) said, "Scratch Monsieur Dumas's hide and you will find the savage . . . a Negro!"[12] I should note here that scratching the peridium (or outer layer) of a black truffle is one way to discern the real deal from "counterfeits"—*Tuber melanosporum* will have a brown layer just underneath, while the less fragrant *Tuber indicum* will show black.

But Dumas had the ability to ether his literary competitors and other rude detractors with the twirl of his pen, the blade of his wit. He famously said to one who had derided his ancestry, "My father was a mulatto, my grandfather was a Negro, and my great-grandfather a monkey. You see, sir, my family starts where yours ends."[13] The great "quadroon" was a cad and could be a windbag, it seems, but most importantly, he understood the truffle's allure. He was of the opinion that to "tell the story of the truffle was to tell the history of world civilization." He also included a recipe for bear paw in his famous *Grand dictionnaire de cuisine*. Trust me, you need to know all this for reasons.

Jasmine lays flat on her stomach, her chin nearly buried in the earth at the base of a scrawny oak tree. She's covered in that fine micaceous layer and it's making her golden. I'm on my hands and knees, on the other side of the oak seedling that was planted about a year ago, trying to dig the way she digs.

"It's okay, you're not going to hurt it," she says, because she sees how shy I am about excavating the lateral roots of the plant.

I believe her but am still wary. Each of the more than 200 trees that surround us set her back $27. They are also, of course, living, and relatively new to the whole "being a tree" thing. Plus, I'm clumsy and aware of my impatience to examine the roots—this will (hopefully) be my first time seeing fungal material under the microscope. Whether the oaks and pines are stressed by our radical meddling or not, collecting these roots will give us important information. When we examine them in search of *Tuber melanosporum* colonization, it will tell Jasmine whether her operation has begun in earnest the five-to-seven-year process of becoming a real, productive truffle farm.

At certain points in history, truffles were poor people's food. They were more abundant when forests were more abundant, and all a would-be hunter needed was to know what trees they like to grow with and how to retrieve them from the retriever. My own relationship to luxury is typical for one who grew up mostly secure but had a single mother who sometimes went without so we could have. We were raised to appreciate quality, but not expect it. I still enjoy *quality*, but grapple with knowing that *luxury* for me means that someone else, down the supply chain, will, under our current system, lack. More than once during my visit to the farm, and especially when we're digging and excavating, I think about *Wolfiporia*, a large black subterranean sclerotium (a fungus's nutrient store for rough times) and wild crop that was dug up and eaten sometimes by people who had self-emancipated from slavery. Sclerotia from fungi in this genus have been used in Chinese medicine for millennia. For those dispossessed on the so-called North American continent, finding the coconut-like, tuber-like mass would mean digging up an immense store of medicinal sustenance. Dried, cubed *Wolfiporia*, procured at the Chinese market down the street from my apartment, has a farinaceous odor. The self-emancipated likely learned where to find the sclerotia from local indigenous tribes.[14] I imagine holding a truffle in one hand and a tuckahoe (a Native word for underground food sources like *Wolfiporia*) in the other, a visual embodiment of the continuum from survival to luxury.[15]

Before I looked through the eyepiece of Jasmine's microscope, the last time I'd touched one was in high school chemistry class, about two decades before. That was post-lunch, and while my teacher was great, there was something about tiny specimens and tweezers and glass slides and solutions and mounts and objectives that felt too finicky for me to take to. Getting it all right seemed very complicated—much less getting it all so right that I saw exactly what I was "meant" to see, and could come to conclusions based on that. I'm similarly unmotivated to take on baking projects that go wrong if your measurements are off by three grains of flour.

So I thought that using a microscope was not for me, until that day with our excavated lateral roots. Whereas in high school, I had little interest in the things we were trying to see more clearly, like blond versus brunette strands of hair or onion skin cells, this would be different. What chemistry class microscopy lacked, for me, was the drama of expectancy. But when Jasmine turned the light on under that root tip, she gave me the gift of the mushroom hunt in miniature—a search that had meaning. We were looking for a sort of "hairy" "glove" around the "finger" of a *Quercus ilex* root. Strangely, I don't remember who looked first, whether Jasmine did the teacherly thing and made me tell her whether the root held what we were looking for or whether she was too impatient to know herself.

Whichever of us it was, we did not see the handiwork of the holy of holies. High school chemistry class never held this kind of dismay, either. My disappointment on Jasmine's behalf reminded me of what I felt when I first tasted a truffle, expecting the manipulations of that cloying, synthetic oil and getting inert matter. The root we examined instead showed evidence it had been colonized by something undesirable. Despite the brief letdown, Jasmine knew it was still very early days. Later, her Spanish mentor reassured her that things would start to happen in due course.

The next morning, I peep through the blinds in the guest bedroom to watch thick fog lazily burning off the stover as crows forage within it. Black birds rooting around makes me want to root around too, but this isn't my house and I don't want to wake anyone with my movements. Plus, I haven't asked Jasmine where exactly her property ends, which would make just wandering risky. I'm feeling a little inner hecticness—it's been a minute since I've strolled under continuous canopy. Are we far north enough, I wonder, to be in *Tuber canaliculatum* country? That's the Appalachian truffle, which grows wild in eastern North America. I learned about it on the morning of The Flavorless Expensive Scrambled Eggs™, vowing that in the next chapter of my life, when I live in a place where I'm able to house a dog companion, I would teach my new friend the scent of this native truffle, which we could maybe find together.

Mycology elder Walt Sturgeon told me that when he transported this obscure truffle, its aroma was so powerful that he couldn't keep it in his hotel room. I want to be overwhelmed in that way, to be knocked on my ass by the hidden, unexpected fruits of creation, just me and a pup. My partner snores where I left him in bed. *There is no separate paradise,* I think, though it's way too early in the morning for all that, *just the one that we ignore and destroy chasing after capital.* A couple of crows in the spent corn look to be tussling. I wonder if this restlessness is how my grandmother felt on the rare occasions she didn't make it to church. Like she hadn't properly informed the Lord that she was grateful that week. I'm grateful in this moment, at my favorite time of day, when it's still all potential, before the minutes button themselves down and the sun etches sameness into place on the landscape. Black birds peck about in the haze, gossiping their morning gossip while, perhaps, a Hartig net minutely stitches itself into being around the subterranean root of a downy oak.

I've come around on cultivation, and Jasmine is the one who brought me round. Not least because what she's doing involves more secrets, and a whole lot more intensity, than your average grow. I find it poetic to see her taking up so much space in a speculative future with the fruit of her knowledge. And in many respects, what she has magicked into being with her consummate practicality and scientific know-how has all the impractical beauty of an art. She's created a small patch of the Mediterranean in Virginia. Under her guiding hand and watchful eye, the land, which has so often had to witness atrocity, nightmare, and labor without beauty, is dreaming. She's come a long way from the little girl on Okinawa who believed balling up flower petals in your hands turned them into seeds you could plant.

In the time since my visit, Jasmine has found brûlés around some of her trees, which I hope means it won't be long till the fruits of her labor land on the market. Since that time she has added a new truffle species to the business plan, given talks at major mycological conferences, come to visit me in my new apartment, where she christened the microscope my partner salvaged for me, and begun an apprenticeship with a cousin who has decades of experience working the land. She has fretted about the bear den on her property, alerted her neighbor to the problem, then put out a call on Facebook asking for everyone's favorite bear recipe. As soon as I can find a translation of the one for bear paw in Dumas's *Dictionnaire*, I'll go looking for that post. Maybe the dish will involve the holy of holies.

I'm glad to say my elaborate plan to turn Jasmine into a friend has worked. I'm glad it did so just in time for me to watch her alchemize the soil's potential to grow these heady black agents of hypnosis, that stretch our memories all the way back to our chemical ancestry. I'm glad to get to watch her tuck the land in so it dreams. Already she's taught me to treat the future as a place where the orchard is already flourishing, where my roots are bearing fruit. It's a lesson I don't take for granted, since tomorrow's arrival has rarely been part of my plan. A forager and a farmer aren't always bound to share the same view; as in all things, Jasmine's way of looking at the world and at money are consummately practical and founded on the here and now. I have one foot already in the "big one," the agglomeration of crises, the coming apocalypses. But our two approaches need each other, and their coming together is holy. We once argued into the night about how the pursuit of money does and does not translate to a practice of freedom. Or, at least, that's how I now understand the tension between the things we said, which by turns dissolved and merged and went chimera as we drank wine and laughed and got pensive and sighed and borrowed against tomorrow.

IN SEARCH OF

Ophiocordyceps unilateralis

(THE ZOMBIE-ANT FUNGUS)

On eusociality, the passage from zombi to zombie, whose clock is inside us?, dissociation, the death drive

To be a person is to be worried that you might not be one.

—Timothy Morton, *Dark Ecology*

Coherence is when the parts of an object weirdly overlap so that they become the "same" thing . . .

—Timothy Morton, *Dark Ecology*

The ant lists this way and that, like she's cross-faded. She has never been this off task. Anytime she tries to ascend into the canopy where her life has been spent, in galleries carved through the dead branches of the tree in service of her queen, she convulses, staggers, and falls to the ground, where she's pulled from one queendom to another. Where her volition was—or where the things that might be considered her volition would be if she was truly individual, rather than a cell in a superorganism—there is now *Ophiocordyceps unilateralis*. It works through her, straining her corpse toward an undreamt-of goal. Is there anything more lonely than this, a eusocial being in her moment of mortal digression away from the body of the colony? Anything more supremely wrong? She is a zombie.

This is the technical term. This is how the biologists refer to her, because "parasitized ant" doesn't contain the cross-species body horror an onlooker is sure to feel when, at solar noon, the parasite makes the ant clamp violently down on a vein on the underside of a low-slung leaf and stay clamped there, until well after her eyes have gone dim. She looks like herself, but she is mycelium made ambulatory, until such a time as her body more directly expresses its new genetic aims by becoming a fungal sculpture, bursting cottony at the seams. And what a sculpture it will be: a little too much, self-consciously macabre, a bit emo, to be honest. How like her brain on a stake the perithecia looks, as if speared by this new horn on her head. With such ornamentation, the shell of her frozen in a stance she would never have taken in life, she is a former ant, teeming with her own death.

I was seven or eight. One day, in the gap between walking home from the bus stop, and an adult coming to rescue me from self-regulation, I remember pressing the skin between my mouth and nose right up against our back screen door. I was not allowed to be outside post-locking-the-door-behind-me, pre-*Frasier*. What if I staggered into the pond or fell from a tree or was absorbed by the saw palmettos or bumped into the grandfather clock, stumbling into another dimension? My mother didn't want the headache. This latchkey time alone (and free) was to be hidden from unfriendly, nosy neighbors; a surveilling force we couldn't afford to pique.

Inasmuch as the person I am now is continuous with the person I was then, that indoor time alone was probably an attractive wilderness until it wasn't, until I noticed the walls morphing into seams I wanted to burst from. Kid sadness is the worst. It can be so energetic. I held my face to the locked screen door and moved it from side to side, gawping at the chickens, geese, goats, and, some years, peacocks through all the little squares of the mesh like a compound-eyed creature. I moved my head side to side, at first maybe scratching an itch. I continued the motion—was I playing some beat in my head with that rote movement? Some primeval spell, some rhythmic recipe for leaving the body? I moved my head side to side and I moved my head side to side and I did it again like a metronome until the skin in my philtrum, that indentation just below the septum, had been rubbed pink. And I kept going until it bled.

Is it only some of us who teem with our own death from the day we're born? Was this the first time I went looking under my skin for the clock that didn't belong there? Or just the first time I remember?

I had seen kids on television use a magnifying glass to concentrate the sun's rays so that it burned the ant on the sidewalk. In my dream, I follow the ant on hands and knees just before infection, ten days before that death grip, with microscope vision. The spores land on her cuticle and begin to press their way into her musculature. She grooms herself, fastidious as a cat. I say, *Oh, sweetie, the damage is already done.* The day before, she had caressed a different insect with her antennae, coaxing the treehopper to favor her with a taste of his honeydew. In her short life she excavated the galleries of her home to further colonize the tree, she hunted for food, or she tended the larvae, the brood. She was an expression of the fitness of her nest, which makes it easy to appreciate the specificity, the ecology, and the nothing-personal of her demise.

The first time I step out of my body and float away from it, I am neon-bicycle-shorts-with-a-ruffled-skirt-young. I don't remember who my body was with, or even what it was doing, but I do remember the sudden, visceral certainty that I didn't know them, had never really met them, and (though I wouldn't have thought it in these words) that who they were to me was a projection that could evaporate if I ceased projecting, which my brain was in the process of doing—ceasing—without prompting from me. In the years since I have wondered what precisely steps and floats away? What is it made of, what material? *O. unilateralis* spores float away on the slightest breeze; how often have I lain empty, a ready receptacle? What calculus had gone into that vital thing's decision to step away and make me watch myself? Why did the observing feel different from that which is done in the mirror? And the biggest question—why, when most of the time the thing is content to take just a quick breather outside my body, did it sometimes abscond with my self for days, months, weeks at a time? That's when I got into trouble.

Given this tendency, you would be surprised, I think, if I'd never been suicidal.

Littered throughout the journals I have irreligiously kept over the years are lists that look like this:

> w loves me
>
> many homes if i want them, including the woods
>
> mostly healthy*
>
> get to do things i love sometimes
>
> good family

have some friends (real ones)

mushrooms exist

I've speculated: the animal in me gets sick of being trapped inside a vessel that only ever sits, plays, sleeps, thinks, reads, and weeps inside the dying whale of capitalism, an unnatural clock ticking unnaturally loud as the background of the everyday. Because it knows of other, preferable ways an essence can occupy itself and here I am, mostly wasting time. While it goes away to do god-knows-what, it learns over the years to power my body from a remove, to make it shower and eat so what's left of me won't offend or worry or hurt anyone's feelings. I've called this "dissociation," though I don't think that's what a professional knower of these things would do (the ones I've visited usually only looked at me hard; until all too recently they never told me what was wrong [capitalism]), and I've called myself depressed when I got used to having a part of me gone, and what was left of me (in bed, on the ground, in the tub) completely sapped of strength.

In college, while dissociating weekly, I took literature and history classes with names like "Creating the Transnational Caribbean" and "New World Revolutions." During these blocks, my life, which had been afloat in time before, was given historical anchors and context, and I gained perspective that would have been impossible without that academic remove. My mother, I learned, was part of a wave of immigration from Jamaica that was now legible to some as a decades-long "brain drain" (terrible term), imperialism that obligates its subject to move away from home. I read about Haiti and understood why the nation, far more than a simple repository of our childhood anti-Blackness and received self-loathing, was a renegade, a global actor whose attempts at self-determination would be punished throughout its existence; that the people of Haiti would be right to burn the rest of the world to the ground. My mother, our family, Haiti, my self—as Caribbeans we all existed and exist at several hostile intersections, making maneuverability difficult. Difficult but not impossible. It is no wonder then that the zombie, the restless embodied dead, more potent than, more confrontational than, a mere wispy ghost, would be born in this place.

Zombies come from seventeenth-century Saint-Domingue, the French colony that would become Haiti and the Dominican Republic. In this iteration, *zombi* refers to the bodies of those not allowed to pass over to *lan guinée*, literally Guinea or West Africa, that lush green home, paradise.[1] A zombified body can be animated with a will not its own, to pay off a debt. Out of the brutal atmosphere in which an enslaved plantation worker led their life sprang the creature who, even after having died, perhaps by their own hand, must still toil endlessly, the balm of rest an impossibility. In this context, the zombie is a spiritual vessel for unspeakable anguish. A repository of hopelessness. No peace even in death.

Later, after the revolution, what it means to be a zombie will shift—zombification will be the result of a voodoo practitioner entrapping the soul of the recently dead, giving the corpse a whiff of that soul, then pressing the resulting creature into service. This manipulation of the boundary between life and death, which we in the West take as the ultimate truth and an important site of Cartesian rationality, is one reason that the Haitian zombie is such a potent figure. If the world seems to have fixed us in our helplessness, but death is not a fixed state, then what else is possible, after?

Armed with a jeweler's loupe I venture into my home woods and look for Gibellula-like fungi, the ones that transform spiders from cunning stop-motion tricksters to hollow, fleece-sweatered cadavers. The uniformity of the resultant arthropod final resting places suggests these fungi may affect the spider's behavior just like *O. unilateralis* does that of the ant: that is, piloting them to the perfect environment for the fungus to thrive. I have found these small but spectacular mycelial fireworks stitched to the bark of trees several times. It does not get old. After my most recent discovery of one such ascocarp on a chilly November day, I apologized to the tree that held it and the fungus who might not have finished sporulating and peeled the bark to bring *Gibellula* (or similar) home for closer study. The minute fuzz reminded me of a mycoacquaintance and biotechnologist who once documented his use of *Beauveria* to infect house flies at his place. Pest control for nerds. Having oohed and aahed through my macro lens over the many tiny lavender clubs extending from the yellowed abdomen, and thinking of Anansi, I went to the biodiversity cataloguing app iNaturalist. I wanted to look at the program's map and see which cordycipitoid species were being recorded from the Caribbean. Noting the few *Ophiocordyceps* observations in Jamaica, I shifted my attention to Haiti. The only entomopathogen that has been recorded on the app from that side of the landmass is *Moelleriella epiphylla,* a scale insect parasite. Anansi, who goes by the name of Ti Malice in *Ayiti,* laughs as I pin my former spider to a board and mount the cadaver under glass.

In 1937, Zora Neale Hurston went to Haiti and Jamaica on a Guggenheim fellowship to study spirituality and magic. In *Tell My Horse,* her book about the trip, Hurston writes, "What is the whole truth and nothing else but the truth about Zombies? I do not know, but I know that I saw the broken remnant, relic, or refuse of Felicia Felix-Mentor in a hospital yard."[2] To accompany these words, we get a haunting black and white photograph of that "remnant, relic, or refuse"—a thin figure in an amorphous dress with an illegible face and lightless eyes, arms flung back as if she's shambling towards the photographer. The space between these two sentences of Hurston's is expansive; it contains countless possible worlds. Felicia died in 1907 and was buried. In 1936, the "remnant, relic, or refuse" turned up naked on a farm, insisting that she used to live there. Hurston reports that Felicia's remnant is an outlier because, in general, recovered zombies had lost the power of speech. Felicia's brother recognized her body, and it was sent, still animate, to the hospital.

The twenty-nine years between her funeral and her hospitalization haunt me. Where did she go? What was she made of?

Hurston learned that the *houngan,* either hired to seek revenge or working for himself, knew there was an important step that could not be skipped lest the *zombi*'s enslavement prove incomplete: "First, he is carried past the house where he lived. *This* is always done. *Must* be. If the victim were not taken past his former house, later on he would recognize it and return. But once he is taken past, it is gone from his consciousness forever." Whoever turned Felicia into a zombie may have skipped this step.[3]

Zombie ants do not go home.

There are astonishments that hold you still and whisper-shout that earth is also like *this*—that remind you every inexplicable process taking place inside an instant on our planet laughs darkly at our ennui, our nihilism, our death drive. My latest astonishment is *Ophiocordyceps*'s complex and audacious behavior. Even if science is explaining the biochemical and mechanical processes behind *O. u.*'s puppeteering, still: The sheer intimacy of parasite and host? What that clumsy waltz has taught and untaught me about evolution?—that smacks of the kind of secret we will dwell in all our lives.

Here is a brainless being without a nervous system (that machinery we call a requirement for intelligence) that has so completely commandeered a "more elaborate" being's body that that body can be considered part of the fungus's "extended phenotype"—that is, effectively a continuation of the fungus's own genetics.[4] And, researchers suggest, it may be the ant's circadian rhythms (its sense of timing, a comedian's greatest tool) that are "hijacked" by *Ophiocordyceps* to allow this feat.[5] When does she cease to be an ant and become mycelium with legs? The minute her former neighbor releases the spore that will connect with her? When the spore gets in? When she performs the first act that cannot be read as ant instinct? When she stops living on her time and starts living on the parasite's?

Many respond to this dance between insect and fungus with empathy for the former and disgust and fear of the latter. Natural enough, of course. Like ants, we've got a brain and a nervous system, which have pride of place in the Western study of biology. But this focus has made us ignore the vitality and volitions that make up each individual "us," and which may have more impact on who we are than we ever dreamed possible. Slowly, slowly, we internalize that the appendix is not merely vestigial but may be an important repository for gut bacteria, which are important for gut health, which is important for brain health.[6] The insect pathogens have held a mirror up to my body and asked me to look beyond what my naked eye can see, to the rest of my ecology.

The mini-documentaries that describe entomopathogens do so with a flair for the dramatic—as soon as the word "spore" is uttered, dissonant music starts up in the background or replaces the soundtrack of soothing rainforest sounds to let you know the dread villain has arrived to plague the individual hero. The parasite is no good for that "individual" ant, of course (which is of course no way to think about the *Camponotus* species in question because they are eusocial), but is it good for Amazonian ecology to keep carpenter ant populations in check? And is what's good for Amazonian ecology good for *Camponotus* genetics on a whole?

Maybe it's because I learned about cordycipitoid fungi through a mycophilic lens, but "summit illness" and "infection" have always seemed insufficient to describe the phenomena, even though *Ophiocordyceps* is a parasite, its expression a form of insect *pathology* that is no good for the ant. Maybe the keyword is balance? Bacteria become disease when balance is absent. We should sing out "My name is Legion, for we are many," but sometimes one voice drowns out the rest. We are chimeras who are loath to admit it. Multiplicity's the ultimate preexisting condition.

This parasite, this process, raises more questions than it answers, but at my back I hear time's gas-guzzling chariot and still *O. unilateralis* asks me one more question: What have the many organisms that amount to you done lately to know they—to know you—are free? *O. unilateralis* wants to know how I would know whether my movements were my own or not.

Kyle Bishop, an associate professor of English, seeks to put the zombie ant in historical context by reckoning with the zombie's Haitian roots. He also asks what we can learn from the "real-life zombie"—that is, the ant. "To me what the zombie is, is the loss of agency," says Bishop. "Ultimately it's a slave metaphor.

Nothing is more terrifying than losing your ability to control your own body or make your own decisions."[7]

If zombies were only a metaphor for the terror of losing your autonomy, then we'd be awash in them. Losing your ability to control your own body and make your own decisions is the daily lived reality of prisoners, Black people, those who can reproduce, trans people, the colonized, the disabled, the elderly, children, and the global poor, to name a few. If a zombie is a mere metaphor, then what of Felicia Felix-Mentor's twenty-nine years? Who is defining real life here? What world is he looking at as he makes that definitional decision? The author suggests that the infected ant is the real-life zombie in an instructive inversion—the borrowed word once again gets unstuck from culture, unstuck in time. We are the metaphors for them.

One student of *Ophiocordyceps unilateralis* used a fascinating word to describe the parasites' manipulation of hosts. According to biologist David Hughes, roughly 50 percent of earth's biomass is parasitic, yet only a tiny minority of those parasites manipulate behavior, because the adaptation is "extremely *expensive* biologically. . . . Most parasites can do a really good job [of transmitting their genes from one host to another] without having to control behavior."[8] So what makes them do it? Why have they evolved to be so danged ruthless, so danged spectacular? The simple answer is that the ants with which they coevolved would not tolerate the scent or drunken behavior of an *O. unilateralis*-infected ant among their ranks, even if the ant had been able to find their way home. Without the reverse-homing effect of the fungus's will as expressed by the ant's movement, *O. unilateralis* would not have an opportunity to form its fruiting body and to sporulate—the uninfected ants would tear the zombified ant limb from limb or toss them in rubbish receptacles as soon as they got wind. So the evolutionary best bet for the fungus, it turned out, was to isolate, isolate, and isolate again, till such a time as the spores could fly free. Zombie ants do not go home.

When an enslaved African took their own life on a French plantation in Saint-Domingue, it cost the person who called himself her owner a pretty penny. For this reason, overseers threatened their charges with the zombification of their corpses and continued labor. Suicide has been expensive for as long as capitalism has existed. I think about those dead who, in dying, saddled their loved ones with end-of-life costs. On the CDC's website, I find a page about suicide prevention with a purple box headed "Economic Burden." It reads, "The economic toll of suicide on society is immense. Suicides and suicide attempts cost the nation almost $70 billion per year in lifetime medical and *work-loss costs*" (emphasis mine).[9] "Medical and work-loss costs" is hyperlinked. I decide that's enough reading for the day and close my laptop before I can click that link.

During a recent apocalypse, I went to pay for the large cartload of groceries for just me (the hope was that they would last for at least a couple of months, during which I wouldn't have to step out into the commercial spaces where we feared this apocalypse most), and when I opened my wallet, twigs, moss, and leaves fell out of it. When you are an animal, which you always are despite what some people would have you believe, it's essential not to lose touch with your perfect bruteness. We had lost touch. This apocalypse reminded us that our stomachs need feeding and our asses need wiping and that being outside our square homes and rectangular screens is the best way to try to stay sane. My apocalypse jaunts to the woods, during which I would squat to pee and eat invasive garlic mustard leaves right from the ground and raw amber jelly roll fungi right off sticks—the jaunts, in their roundness, had taken over that wallet with its plastic squares, its insistence that I was more than an animal. Seeing those twigs, leaves, and moss on the waxed grocery store floor made my eyes well. Sometimes the apocalypse disrupts zombification. Sometimes it allows us space to hear capitalism's ticking clock, excise it from under our skin.

Over the past decade, headlines like "Get Ready for Sex-Crazed Zombie Cicadas Known as 'Flying Saltshakers of Death'" (*New York Post*) and "How the Zombie Fungus Takes Over Ants' Bodies to Control Their Minds" (*The Atlantic*) have popped up in the news now and again. They use that *z* word as a shorthand for the uncanniness of body-snatching parasitism.[10] I wonder how many of the biologists and, subsequently, how many of the journalists who cite their studies on fungal parasites know the origins of the zombie.

Despite having extremely specific sociopolitical and spiritual origins, the zombie that is known, loved, and spun into many a Western yarn is largely divorced from its heritage. How does the zombie morph when it's whitewashed? In his article "The Tragic, Forgotten History of Zombies," which appeared in *The Atlantic* in 2015, Mike Mariani suggests that, as a result of the transformation of the zombie in the Western imaginary, we get *The Walking Dead,* a show that takes place in an idyllic apocalyptic landscape in which "the importance of the lives of [the non-zombie] characters . . . is implicit, because theirs is the only story left to tell. And that, of course, is the key to their fantasist power: who wouldn't want to escape into characters leading lives of infallible significance, with their survival and the endurance of the human race perpetually at stake."[11] I'm reminded of manifest destiny, a machine whose engine was property and God's will, which generated apocalypses wherever it went. I'm reminded that for the most part, this zombie is an Other, over whom the hero must prevail, even when it has taken over the body of someone known to the hero, perhaps even loved by him. In this mode, a zombie for whom one has sympathy is even more dangerous. The hero looks into its eyes and sees something he recognizes as humanity, but ultimately that zombie wants to eat his brain, to turn him into something monstrous. The zombie-as-other brings on the apocalypse and their role ends there. Nothing fruits from this zombie's life. In seventeenth-century Haiti, zombies are a response to an apocalypse; a reminder that death is not the end.

When I ask biologist João Araújo what he thought of the zombie movie *The Girl with All the Gifts,* an adaptation of the eponymous book and a surprisingly watchable treatment of what would happen if *O. unilateralis* were to begin infecting humans, he says he hasn't seen it.[12] I'm taken aback—he acted as a consultant for the movie, describing for the filmmakers what the fruiting bodies of the fungus should look like. He explains that he probably *should* see it, and will eventually, but that he's more of a documentary guy. Later in the conversation he mentions the BBC documentary that blew his mind with its time-lapse footage of *O. unilateralis*. He'd seen it back before starting on the career path that led him to become the assistant curator of mycology at the New York Botanical Garden. The documentary in question is the same one from which I posted a clip to my Facebook timeline in 2010 with the breathless caption "extremely disturbing and beautiful."

Araújo clearly loves his work—he describes it as a dream job. As an undergrad he set out to study insects with the ambition of becoming an entomologist, but he had his head fully turned when he learned about the fungal parasites that zombify insects. I tell Araújo that he should see the movie that he helped make, if only because, astonishingly, it centers a fungus that threatens to eradicate humanity. We're rooting for *Ophiocordyceps* and its human symbiont to release the spores that will rid the world of all but one human. Araújo says this is interesting, because when people first learn of *O. unilateralis*, a response he often hears is "Poor ant!" (My partner thinks that maybe I was the only one rooting for the zombie, but that can't be right.)

When I google "how many ants are in the world," a pest control website tells me that myrmecologists estimate there are 1 quadrillion. The site must be old, because farther down the list of results, the number is more like 10 quintillion, or 1.2 billion ants per person. The lengths to which we'll go to sympathize with an individual (who is not even really an individual) and be ambivalent toward a group are wild. The number of how-to videos for obliterating an unwanted ant colony on YouTube is likewise staggering. I tell Araújo about the origins of zombies and suggest that it's all about humans being unable to handle that which bears an uncanny resemblance to a human (or ant) but then does not act to forward human will or desires. Few things disturb us more than a soulless "thing" with our form.

Watching *The Girl with All the Gifts,* I immediately pour myself into Melanie. I mentally admonish the other characters to protect her at all costs. She raises her hand with the correct answer, she's eager to please, she's the teacher's favorite, and it's hard to believe but she'll eat you alive. That's what she did to her mother while in utero. When we first meet her, she and her peers lead the lives of prisoners lying in wait, the clock their enemy. Strapped into wheelchairs, they're pushed from cell to classroom, where we grow to understand that their capacities are being tested in this school-like setting and the unlucky ones are sacrificed, one by one, to scientific experimentation. In this universe, the zombie apocalypse has come for humanity in the form of a fictionalized *Ophiocordyceps* that turns people into fast-moving, mindless zombies, but then makes intelligent, thoughtful zombies out of the second generation—those who eat their way out of their infected mothers. What sets this story apart is that Melanie would eat the human characters if they didn't treat their skin with a scent-effacing chemical, and yet we very much wish her to survive. A little Black girl (or more than girl) with the power (realized in the end) to colonize the world.

In the second sentence of the book that the film is adapted from, it is explained that the protagonist's name, Melanie, means "'the black girl' from an ancient Greek word, but her skin is actually very fair, so she thinks maybe it's not a good name for her."[13] She would rather be called Pandora. I wonder why the filmmakers decided to cast Sennia Nanua, who's Black, in the role. In an interview, the young actress said that though she had auditioned several times for the part, the description of the character in the book made her think that she didn't have a chance.

In the book, Melanie observes her classmates seizing the opportunity to ask about their origins, which they don't really know, when their favorite teacher is in a particularly open mood. Of her classmates (and also herself), she says, "The one thing they never learn about, really, is themselves."[14] I like to think the filmmakers simply couldn't let that line be said by a little white girl.

One movie poster shows Melanie's character looming over a ruined post-apocalyptic landscape with a muzzle over her bloody face (she has just eaten a cat). The muzzle is reminiscent of the ones enslaved Africans would be made to wear as punishment so that they couldn't eat. The first portrayal of a zombie in the West must specify that it's a twist on the Haitian myth with the title *White Zombie* (1932). Decades later, director George Romero invented the modern zombie genre with 1968's *Night of the Living Dead*. Romero originally assumed the lead—Ben, the most competent of the zombie slayers—would be white. When Black actor Duane Jones auditioned, however, it was clear that he was the right choice.[15]

Perhaps these casting moves indicate that when a zombie story is thoughtfully told, it is also inextricable from the Black uncanny.

In *The Girl with All the Gifts*, Melanie says that "she knows that when she grows up, she'll be beautiful, with princes falling over themselves to climb her tower and rescue her. Assuming, of course, that she has a tower."[16] When Melanie eventually initiates the reign of the zombie-fungus-human hybrid by setting fire to a towering, spore-bearing fruit body of the fictionalized *O. unilateralis*, it turns out that she needs no rescuing. But we might.

In Manhattan's Billionaires' Row, at 111 West 57th Street, the residential Steinway Tower is over a quarter of a mile high and only sixty feet wide, scraping the sky at eighty-five stories. There are sixty units in the building, each spanning one or two whole floors, with no street-level windows to smash, no neighbors side-on to you, no need to see other people at all, unless you're stationed at the window watching the ants scatter below. The tower's width-to-height ratio is an astonishing 1:24. It's deadly skinny, a feat of architectural design for those who desire supertall repositories for generational wealth, many of which stand half-empty in the middle of the country's most densely populated city—this is Zombie Urbanism.

That's the term used by professor of architecture Matthew Soules in his book *Icebergs, Zombies, and the Ultra-Thin*.[17] The demand for expensive luxury residences that tower in densely populated cities has, in his view, mutated the way we build and the reasons we're building. Learning this, I felt deep sadness at the thought of the ultrawealthy hoarding homes in the sky as they might own a Picasso while, down below, the unhoused are harrowed out of subway systems in the name of public safety like so much refuse. (A recent *New York Times* article about clearing 239 homeless encampments said that "sanitation workers and police" partnered for the effort.[18] Over half of New York City's unhoused are Black.[19] Where do they go?). I feel an arsonist's rage when I learn of a zoning loophole used by the developers of One57, another supertall Billionaires' Row property, which allowed them to claim a $65.6 million tax break when they spent $5.9 million to build 66 affordable apartments in the Bronx. According to the Independent Budget Office, the amount those developers saved on taxes could have built 367 affordable units.[20]

If I shut my eyes I can see two time-lapses throbbing side by side: the erection of a super-slender tower—a dead spot amidst pulsing life—and the emergence of a single slim stroma from a dull-eyed ant's head. Come, Melanie, you who wanted to be called Pandora—show us how to wield that world-ending torch.

Death as a possibility, as a vessel for another life. Zombie as that which lies suspended until the next world can be born. Our bodies do not rot here, they remain uncorrupted until they can be useful again.

Professor of English Jacob L. Bender, in a book called *Modern Death in Irish and Latin American Literature*, suggests that the zombie transformations in Alejo Carpentier's *The Kingdom of This World* are liberatory.[21] He quotes another text, "The Zombi/e Manifesto," which says, "The Zombie is currently understood as simultaneously powerless and powerful, slave and slave rebellion." Bender argues that what often gets lost in discussions of the zombie is that in some sources, as anthropologist Melville J. Herskovits notes in his 1939 book *Life in the Haitian Valley*, the zombified are not just the animate dead but also bodies capable of metamorphosis. Herskovitz quotes a certain Dr. Parsons, who notes that "a ganga [or sorcerer] who keeps a zombi may transform him into a stone, perhaps in front of his house, or into any kind of animal. Transforming him into a pig or sheep or cow."[22]

In *The Kingdom of This World,* formerly enslaved Ti Noël transforms himself into numerous animals, waiting out the continued enslavement of his people under the presidency of Henri Christophe:

> Once he had come to this decision, Ti Noël was astonished at how easy it is to turn into an animal when one has the necessary powers. In proof of this he climbed a tree, willed himself to become a bird, and instantly became a bird. He watched the Surveyors from the top of a branch, digging his beak into the violated flesh of a medlar. The next day he willed himself to a stallion, and he was a stallion, but he had to run off as fast as he could from a mulatto who tried to lasso him and geld him with a kitchen knife. He turned himself into a wasp, but he soon tired of the monotonous geometry of wax constructions. He made the mistake of becoming an ant, only to find himself carrying heavy loads over interminable paths under the vigilance of big-headed ants who reminded him unpleasantly of Lenormand de Mézy's overseers, Henry Christophe's guards, and the mulattoes of today.[23]

The first time I break the casing of a Bic lady's disposable razor and press the metal into my wrist I am fifteen. "Parasuicide" is sometimes described as non-fatal self-injurious behavior with a clear intent to cause bodily harm or death. The word "clear" gets stuck in my throat. Para—"alongside" suicide? My death travels with me. The spore of it. If I can get over the aversion to the corruption of my own flesh, it's always there, alongside. Look, I don't know why I was that impatient to see for myself whether death was inside me, whether I could dig deep enough to find this thing everyone was obsessed with and nobody talked about, but I was. Call it youth—you stay young for as long as your curiosity is boundless. Or you stay young if you die that way. Dissociation and depression and self-harm are all supposed to be constellations of the same dysfunction, but I've cut my skin on a day of delirious happiness, and sometimes it seems to me like I want to take that happy feeling with me to some other place.

"Do you like yourself?" my mother asked once. "It doesn't seem like you do."

I answered *yes*, though later, in my room, I turned the question over in my head until I cried.

The right question, which neither of us could have known at the time, was, "Where did you go?"

Now, on the verge of leaving my thirties, the correct questions have become, "By whose clock are you living your life?" And, "What are you and all the organisms that hold you together doing to be worthy of your freedom, to practice it every day? To know that your movements are your own?"

At the end of our conversation, I ask Araújo what species of entomopathogenic fungi, besides scale insect parasites and *Cordyceps militaris*, I might expect to find this far north in Vermont, which is where I am while we talk through Zoom. He tells me that *Gibellula* can be found up here on the undersides of leaves. "So just turn thousands of leaves, and you will find them."

Just keep turning leaf after leaf after leaf.

IN SEARCH OF

Amanita phalloides

(THE DEATH CAP)

On telling new origin stories, the relativity of poison and of cures, poisonous education

CURE

When the ships pushed back from shore, there were already spores nestled in their sails. The spores teemed against the woven linen, seething with future. Their forebears had sent them soaring on cool, polyphonic breezes from gills and pores and teeth and cups; the whistling wind told them where and how to live and what to eat. The wind also held memories of home.

These Spores were germs of breathing sculptures, of labor and movement in obscurity, of the souls of strange trees and grasses, of rock eaters, of bug hijackers, of flesh that savors like an animal's, of medicine. In a few, the knowledge of all sources of pain and the abolition of time. In all, devices to merge and entangle. The trip was long, and the Spores did some forgetting along the way.

But they observed the oceans and relearned earthly forms to take and ways to make their fruitings dance on invisible currents, on loops and hiccups in time. From the ocean floor, they learned the paradox of luminescence and secrets kept. From feeding shoals of fish, cyclonic grace. From seabirds, the forms of wings. From whale falls, the breakdown of the dead.

Ashore again, they surveyed for new homes, borne inland on the sun-loved skin of their kin. Their kin's work was never done. Their kin's backs bent to breaking. Their kin's limbs were hacked by mute machinery. Their kin were forced to singlemind the land with sugar. All were darkly mutant in this old place with a new name.

But the kin, in their whip-tortured new tongue, ingenious, schemed freedom.

Meanwhile the Spores spilled over, fingering their way underground, stretching and climbing and pulsing with plant, animal, fire, water, stone, and air intelligences. They made silver weave that never dies. They grew with sugar cane and from ackee trunks like ears, like purple eyes from dung. When the earth lapped human blood, the weave learned, again, how to poison, to stop a heart.

This poison moved through kitchens so that Big Men died. The Big Men died and died and died again. The poison weave helped break Big Men down small beyond their meat and smallest critters. Until Big Men went dark. The weave went dark, too, for the most part. The poison is so big now, and the weave so dark, poison big as the Big Men who went back to their cold home, weave as dark as Big Men's coldest night, that the kin can't discern what from what. What is poison? What are the true sources of pain? Where is time suspended?

Who will walk barefoot in the bush to hear these answers?

"Pharmakon," in philosophy and critical theory, is a composite of three meanings: remedy, poison, and scapegoat. These meanings derive from Greek originals: the first and second senses from the Greek φάρμακον (phármakon), denoting any drug, and the third sense from φαρμακός (pharmakós), a ritual in which a human was exiled or sacrificed.[1]

"Poison, poisoning, poisoner!" must run through the minds of some people when I'm introduced to them as a mushroom hunter, given how they react. The arched eyebrow. The up-and-down look. The confession that they "could never," subtext being that they value their life too much to entrust it to the scheming chemistry of such a dreadful kingdom. It's odd, but they tend to adopt this attitude well before they learn what my particular hunting practice looks like. This prejudice does not display itself when I'm introduced with the nonspecific "forager," though eating the wrong plant is no less dangerous. You and I don't need to kid ourselves about what else might be at play when a certain set looks at me and hears the word "hunter." Sometimes, perhaps, during this introductory interlude, they have wanted to express admiration that I've learned enough to eat wild mushrooms, but what comes out often feels like thinly veiled distrust, as if I've been revealed to them as a dabbler in dark magic. The response used to bother me. After all, so many dedicated people work and have worked to change the North American attitude that "mushroom" is synonymous with "poison," and that, unless they come undercooked on a delivery pizza, fungi only deal death. It used to bother me, but now I kind of like it.

During that quiet tussle with their judgment, I fall into a fascinating lineage, some far and unaccustomed loops of history, corners I've only recently begun to explore. The ones where we were poisoners, potioners. Poisons, toxins, adulterants, pollutants, and cures can define who and what we are, and who and what we aren't. To be a definer of these categories is an absorbing job, and there are powerful forces all over the planet with a stake in who and what does that defining.

For some, mushrooms embody the word "poison" (both noun and verb forms). That mushrooms should exist near to where we do our living, with all that cell-destroying potential at the ready in their flesh, without telling us everything about them all at once, is enough to indict. They are always already guilty. Heh, I know that feeling. It is an odd logic—the same one that results in the preemptive killing of a venomous, or nonvenomous, snake who has happened into one's yard. The same logic that hangs a witch.

You don't really get to pick your poison at first—the one who will give birth to you does that. If they can resist, they'll make sure they don't ingest certain things when pregnant and nursing. This is the last time they'll exert as much physical control over what you take in, to say nothing of the psychological. After you're born, whoever takes care of you might slap the mushroom out of your hand when you're in your grazing stage (the age when you're most likely to be ministered to by poison control). Later, they maybe shake their head when they see how much sugar is in the muffins that come with your school lunch, but if they ever had a say in what gets put in your body, this is when they really lose that. They tell you about moderation before your immoderately hormonal body and developing brain can really know what that means. They warn you to stay away from the kids in the neighborhood who make laboratories of their bodies because they want your body to know a kind of purity—they want your baseline to be easiness inside reality. No matter if the reality you've been born into is itself a creeping sort of poison that will sicken if mainlined.

But it's for our guardians to want to shield us from the destructive potential of a straightforward poison.

To learn which mushrooms are used by humans in one's area is to learn just how culturally bound the concept of poison can be. *Those* people in *that* country eat the ones with rocket fuel in their fruit bodies, they just boil them first. Cooking *that* one breaks down its toxicity, but we don't really do that here nor would we recommend it. I just don't see why you would bring *that* one with its trace toxins to the table when there are so many other delicious edibles fruiting at the same time. *That* one looks so incredibly similar to another, deadly species that we don't bother eating it, though it's delicious. People around the world eat mushrooms that look like *that*, but here we don't do it because of a rumor that was started long ago whose truth value was never reassessed. A person's relationship to the poisons they do and don't consume is thoroughly bound up in their heritage and the land they call home. In fact, that relationship is one way groups fortify their sense of self.

I am American; therefore, I dwell in poison. I eat poison, drink poison. I am poison. It's not always toxic masculinity to say, "That girl is poison." You are what you eat. She's all sugar and spice but too much sugar is poison. The dose makes the poison. Warning: Don't forage along the banks of the river in the town you just moved to, the water is poisoned. For nine years, the water of Flint, Michigan, was poison. Couldn't eat the volunteer yard tomatoes at my co-op in the city—the soil sample showed poison. We found out weeks after the chemical fire why the forests of the Blue Hills smelled poison, like plastic. The microplastics in our blood, our lungs, the clouds, the rain, the soil, our food, may be poisoning us. To keep the smell of sweat at bay, talc your breasts, your mound with this poison. If it's too nappy, straighten your hair with lye, a poison. For clear skin, take tetracycline, a poison. To cure the colon damage done by that tetracycline, take mesalazine, a poison. When the lye gives you cancer, start your tango with chemo and await good cell death from the poison. To stay beautiful past your expiration date, inject this harmless poison into your forehead. We found out too late that our online social networks are poison for kids. As COVID raged, we learned our brains alchemized the stress chemical cortisol into poison. Just before any mushroom hunt when the weather's warm, we shellac our skin with a solid coat of DEET, a poison. As reward for finishing this list, I take a sip of wine, my trusty poison of choice. My grandfather, whom I get it from, picked his poison and picked his poison, and picked his poison again.

I am also Jamaican, and I eat poison for breakfast. Stewed with tomato, green and Scotch bonnet peppers, scallion, and salt cod, then served with breadfruit.

It's my first visit to Jamaica during the Christmas holiday in decades, and my last visit anywhere a plane can take you before almost the entire world was grounded from flight. In a few months, a poorly understood agent of death would begin menacing the very air. But before all that, one last gathering to glut ourselves on jackfruit, Tastee patties, and spiked sorrel. Though my mother's side of the family is scattered around the world, its headquarters are my grandmother's house in the St. Andrew Parish hills, and this is where I stay when I'm visiting—my ancestral home. Since my childhood, kitchen and bedroom additions have been built, more than doubling the house's size. The structure's original footprint sits within the present-day living room, and my mother, aunt, and uncles grew up in that tiny space together with my grandmother and grandfather. Though all of the district was poor, my mother's classmates teased her for being the poorest of the poor.

I went back each summer after moving to the States up till my latter teenage years—the back and forth felt like moving between the life I was meant for and the one that would be a readier conduit for the American dollar. In Florida I was odd, awkward, and singular—in Jamaica I was an erstwhile community member, Miss Darling's granddaughter. Back in the day, my cousins and I romped about the property playing dandy shandy, we terrorized the semi-feral cats who scraped for fish bones set out for them on the red clay dirt, we ate our fill of mangoes and guinep, sucking the pulp of salted limes just plucked from the bush, they teased me for my broad vowels and taught me the entries for the year's Independence Festival song competition, I surprised them with my collection of fancy pogs and slammers and by pulling out my fading, wonky Patois for a laugh.

On the walls in the house were photos of cousins and adopted family in graduation gold and black. These days, if I close my eyes, I can still see the ghost of a framed, white Jesus. What survives are wooden wall hangings depicting the Piscean fish and Taurean bull, for my aunt's and mother's astrological signs, respectively, and porcelain tchotchkes my younger brother and I bought from the dollar store for Mamma's and Aunty Nov's Christmas presents. There are new batches of baby pictures on the entertainment center and shoved into the frames of mirrors around the house now, some featuring people I don't even recognize.

In this newish indoor kitchen, a replacement for the outdoor one that was curing her to the bone with woodsmoke and sweat, my ninety-seven-year-old grandmother is teaching me how to ensure I don't kill anyone with the breakfast we're preparing. She oversees while I separate the ackee aril from its seed, careful to dislodge the papery raphe around it, as these bits contain deadly toxins known as hypoglycins A and B.[2] I don't remember who taught me this sacred process before, but I'm pretending to be ignorant of it so that my grandmother will teach me again. The fruits we deal with were allowed to sun ripen, opening up on the tree like woody, three-eyed flowers. Or like one of those aliens who opens their mouth in stereo, if you know what I mean—there are extra mouth flaps or mandibles or whatever. A specimen that has not opened in this way can kill you. The saying goes like this: "Me fadah send me to pick out a wife; tell me to tek only those that smile, fe those that do not smile wi' kill me."

The fruit is fatty; handling its viscera has the feel of butchering an animal. In the phone-camera video I snuck of this lesson, whose focus is mostly my fingers digging their way into that buttery yellow flesh, I can hear my grandmother's steady rasp in the background, telling me how to discern the difference between fruit that is properly ripe and fruit that has gone bad. She is not at all confident, once we get things simmering, that I won't ruin the meal in a matter of minutes with my inattention.

"Take time and stir it, ennuh. And watch it, mind it stick to the bottom of the pot."

Once the dish is fully and properly prepared, the arils resemble miniature yellow lungs among bright veggies in coconut oil dyed orange by tomatoes, those lungs freckled with a generous amount of black pepper. With a starch like breadfruit, green banana, or hard dough bread to sop up the sharp, sweet, spicy, cheesy flavors, I bite into the truest taste of home there is for me. As a keeper of ackee's secrets, my grandmother would have been the first to quell my fear of poison.

The most famous wild mushroom, *Amanita muscaria,* that white-jeweled redhead beloved of children's book illustrators, is an icon among poisons. David Arora called it "the Bloody Mirror of the Galaxy," which puts me in mind of its luminosity, its association with death, and the way it seems to reflect us back at ourselves. Four or five fruits of it are on the menu for me and my friend H on a balmy Bohemian October night at her apartment in the Czech Republic. We're enjoying them in a ramen-inspired soup with scallops and quick-pickled radishes, and they are, against the odds, crisp and meltingly tender. A couple of days prior to our dinner, H took me to the forest for my first international mushroom hunt. When I spotted the *Amanita* fruits, I was starstruck. The red fly agaric's fraternal twin sister grows in the Northeastern US, where I live, but she's a more retiring blonde there, cooler than her sister if no less lovely. This red, though, was so dazzling among the grays and browns of rotted leaves and dried pine needles on the forest floor that I might have screeched. Like you do at the opening notes of your favorite song when you're seeing the artist in concert. It was that deep.

Ask anyone to paint a wild mushroom and this is what you'll likely get; start typing the word "mushroom" in a text box and an emoji version appears. When H sent a photo of our basket of edibles to her mushroom-loving friend, he was aghast, imploring her not to eat anything that I, clearly a murderer or, worse, a dilettante, prepared.

Here's the thing: murder-suicide by mushroom poisoning is not how I would ever plan to go. A mushroom death is not a pretty one—one does not cough lightly into one's kerchief and expire with dignity. But this toadstool can be eaten without intoxication (in the most literal sense of that word), when prepared correctly. The preparation, in our case, included boiling the fruits in seven changes of water until the color leached out, then dry sautéing them to ensure caramelization. Only then was butter added to the pan, where they sizzled until golden brown. When they were prepped this way, my dining partner and I remained unintoxicated but were lightly impressed. Was it worth all the trouble of the water changes? I'm not sure. For me, this process might be more of a party trick. And a lovely measure of a good friend's trust.

And trust was required, to be clear, because even though these mushrooms can be made edible, they contain neurotoxic isoxazole derivatives that can induce "nausea . . . confusion, visual distortion, a feeling of greater strength, delusion, and convulsions."[3] Sometimes patients present at hospital in a comatose state from which they cannot be roused. That's the clinical description of what can happen, though the effects of these mushrooms have a spiritual significance that I won't get into here.

So I understand the reticence with *Amanita muscaria*. Red is a warning color; when other kingdoms of life wear it, it often means stay away. I've heard this reservation expressed on many a foray: "But this one is so bright, isn't it saying 'look but don't touch'?" In this way and in so many others, kingdom fungi doesn't follow the usual rules. I've seen some liver ruiners in my home woods, and by the color metric they looked like snacks: one was so stark white it glowed like the flesh of fresh fish; one had a grayish green cap and greenish chevrons down the length of its stipe like any innocent vegetable. Both were gracefully beautiful, with tissue-thin skirts that came off at the subtlest coaxing. Not a lurid color in sight—just muted, stunning noxiousness.

In some lifetimes, we don't know what the real poison is because we've accepted a story for too long.

When I take an absolute beginner to the woods and the going is good, sometimes the most burning question they have, one they repeat in an "Are we there yet?" tone when we find absolutely anything, I mean it could be half-rotted and smell like a corpse, is "Is this edible?!" Sometimes this is asked even before we've pinned a name on it. If I know what it is and whether it's edible, I'll maybe give my answer, draining my voice of the usual foray leader animation so that they know they're being exhausting. Soon after, I'll slyly segue to The *Amanita* Talk.

"There's a toxin produced in such concentration in mushrooms of this genus," I might say, conversational as all get-out, gesturing at a cute *A. flavoconia* pin or one of the metallic grisettes, for instance, "that, eight to twelve hours after you've sautéed your find in duck fat, it will begin the work of undoing your liver. It will labor at this over the course of a week unless the right medical attention is sought and administered. But first, your intestines will empty themselves through all available exits. Those who've had cholera will be familiar with these symptoms. This is phase one. When this phase is through with you, you'll feel weak and empty, perhaps unable to leave bed. Your body has just done its level best to rid itself of the thing doing its level best to kill you, but it may have met its match." I might pause here and give my audience a once-over, to make sure I've earned the required attention.

"The next step is the honeymoon. Twenty-four hours or so into the geyser stage, the most violent of the symptoms may stop, making you think your unpleasant tangle with food poisoning has finally relented. If you're in the hospital, and the medical professionals don't suspect amatoxins, they might discharge you at this stage. Meanwhile the amanitin has been inhibiting protein synthesis in some body parts that should really continue to synthesize proteins, namely the gut, the liver, and in some cases the kidneys. There may be a recurrence of gastro distress. Necrosis, massive cell death, is not far off.

"The final stage is organ failure. Hepatic encephalopathy and the concomitant confusion, delirium, convulsions, then a coma. And then, it's off to that great mushroom hunt in the sky, where you can presumably eat any mushroom, to your heart's content, without ill effect." I might deliver this last sentence with the evil bitch smile one would expect an evil bitch to smile. Then I might continue:

"There are also perfectly delicious, edible *Amanita* species. I've watched experts disagree about whether a mushroom I'd photographed was tasty *Amanita lavendula* or the most toxic mushroom there is. I'm saying this to say, and believe me I take no pleasure in saying it because I find the aphorism loathsome, that all mushrooms are edible. Some only once."

Hopefully we've been walking through some beautiful forest as this conversation takes place. Hopefully I interrupt myself to point out some cool plant or insect. Hopefully I've made a convert.

In delivering this monologue (or, more likely, a truncated version), my intention is not to spread mycophobia. Far from it. What I hope I'm doing is suggesting that my overeager audience treat these organisms with respect and try to rid themselves of the kind of "I need to put all I survey in my mouth" attitude that has no place in the careful halls of sensible mycophagy.

But the example forager above is not the rule—for the most part, in the youthful, so-called United States, as I've said elsewhere, the attitude toward mushrooms is that one should avoid them at all costs. In fact, in an otherwise beautiful book put out by a folklorist about poisons only a couple of years ago, I found this perplexing sentence: "The only member of the amanita family that is not necessarily fatal is also the one that has become the most recognized of our fungi."[4] (Here she referred to *A. muscaria*). It's odd, our preoccupation with the discrete poisons found in the flesh of a single organism.

The interesting poison is collective.

POISON

When I was fifteen, lonely, isolated, I ate a poisonous amount of over-the-counter sleep aid. In present-day Florida, in the neighborhood where I lived at fifteen, I pluck a vomiter parasol from a lawn of Augustine grass. I hold the mushroom up to my ear and place a call to myself at that age, on the day before I'll push twenty little pink pills from their blister packs and swallow them down with bottled sweet tea. I'm curious to ask this self-obliterative specter with my face, who haunts my memory, why she felt there was no version of the future worth chancing to meet. I know the answer, of course I know the answer, but I would like to hear it in her words (I long to hear her speak—she had a bold way of saying things that frightens me now), and I bet she would like to be asked by someone who knows the answer, someone not asking just to prescribe more pills and not because they must know the answer or be liable.

Maybe I'm too smug in my knowledge that time will take its knee off her neck and start running away from her in every direction at once, relatively soon. And in this possible world, where I am now, which I could give her the coordinates to, she'll want to give chase after time, she'll make a game of bounding through the woods and cutting time off at the pass; she'll actually like trying to tame it. Maybe I'm too comfortable in her future, sipping tea surrounded by a riot of books in which the unofficial story, where she can start to make her home, unfolds. Even though we maybe can't relate, I want to try.

As she nears the vomiter fruit body that I've plucked in her future, which is so much like a brown and white lace umbrella it even has a runner, it tinkles out a breezy melody and drops green spores. She stops dead and cocks her head. She senses there's a moment asking to be born here that will pull her through a portal, blind her with its hospital lights, slap her to get her breathing. She picks the mushroom. She turns it in her hands, looking so interested—like this is true appraisal, like the curiosity will take her past tomorrow—that I know things will be all right, that she won't have to see that look on our mother's face. But she decides it's probably poison.

I'm yelling at her over the years through fungal tissue—Look closer! Hear me!—but she's not accepting messages from this day and age. She throws the mushroom against the side of the neighbor's house, and it reports a satisfying thwack. The next time she'll notice this species, she'll be in her thirties, with scars she doesn't try to hide anymore when she proffers a mushroom for another to inspect. She'll pick it from an Augustine grass island in the parking lot at the beach, and a man with a Weedwacker will tell her not to eat that fairy ring mushroom; it's poison.

The definitions of poison are legion. But here we'll put it simply—a poison is something that, taken in too great a quantity, can harm.

Some poisons, when correctly titrated, are cures. Some adulterant in your food is fine. It's the level of toxin in the soil that you must worry about, veggie-garden-wise. They measure bacteria levels in the swimming and drinking water so you'll know when those are unsafe. There can be safety there. This should soothe the anxiety. Poison is not black and white. I am in love with the gray of it, such beautiful, dissonant gray, I want to backstroke into the middle of it. It is not safe for black and white here, amid poison.

But watch out, don't be too soothed! This one's new: We, on the East Coast, must now pay attention to the air quality to make sure it's all right to breathe outside. Mustn't fall asleep under the stars without checking first. Mustn't hold out your tongue or your drum to catch the rain—for the first time in history, it's no longer potable. There's poison in our food and medicine, deep in the land and wafting in the air, woven into the way we're educated, and we must guard against it when we know it's there. Because these poisons can be odorless, invisible, illegible, and unthinkable.

Ackee is not native to Jamaica. When you google "how did *Blighia sapida* get from Africa to Jamaica," the results all feature a suspiciously similar sentence amounting to its being ferried there "probably aboard a slave ship." This speculation begets speculation—was the smuggler a slaver or the enslaved? Some sources imply that botanist Dr. Thomas Clarke is wholly responsible for introducing the tree to the island, but its initial passage is shrouded enough by the repeated use of the word "probably," in so many sources, and by claims that he "obtained seeds from a West African slave ship," that I choose to remain agnostic on the matter of its arrival in North America.[5] The available sketches of this moment are inactive, and devoid of the Black actors that almost certainly animated it.

Ackee was given that ugly Latin name to honor the British naval captain of the famed *HMS Bounty*, William Bligh (I imagine his one redeeming quality is that he was played by Anthony Hopkins in that one movie about the mutiny), simply because he took seeds or fruits of it to Kew Garden. Perhaps it's assumed by the compilers of this story that we will assume that the ship's science officer (did slave ships have science officers?) is responsible. But Western scientists have rarely been shy about taking absolute credit for the ways they've architected the modern world. And the fact that Captain Bligh is credited with "discovering" the fruit because he took it to a botanical garden is annoying enough. Before him, ackee was "unknown to science." I want it explained to me how a plant that people were already eating and making use of in myriad other ways, brought to and established in "the New World" because of this, and because it would grow luxuriously, could have been "unknown to science."[6] What a narrow view of the countless avenues for humans to know. I want to caper about in the silences and omissions and lost records riddling the official story, to luxuriate inside their expanses.

Which I suppose means I want to be much more than agnostic, envisioning a woman in that hold, maybe my extremely great grand-auntie, having braided ackee seeds into her niece's hair ahead of the voyage, her canerows (which is what we used to call cornrows) teeming with future.

I don't seek out Transatlantic Slave Trade or Slavery Times™ narratives—not to tell as a writer, not to consume. They simply arrive on my plate whenever I try to learn about the movement through time of the people and things that interest me. Thank goodness I've always held those tales at a distance, or the despair might well have swallowed me like molasses. It's easy to overdose on them because they're a background process to absolutely everything here, and we see them and see them and eat them and drink them without much present-day reckoning, let alone meaningful gestures at repair. I've watched the resulting illness ravage bodies and minds. I'm impatient for us to be done with the accounting and move on to the repair. The endless accounting. The nasty lie of the official story.

But I'm Black, so the official story has never been of much use to me anyhow. If we want to survive—those of us who have felt rootlessness without the possibility of future, who have considered suicide or done more than consider—we have had to build alternative pasts for our futures to live in. Those stories are marginal and extralegal and anarchic and mycelial. You cannot kill them in a way that matters.

What happens when old stories teem with new futures? When I look to the likes of Alexis Pauline Gumbs and Tracy K. Smith and Fannie Lou Hamer and Minetta (see below) and Octavia Butler and Saidiya Hartman and Jamaica Kincaid and June Jordan and Fred Moten and James Baldwin—the list goes on—I see the building of our new cosmologies is well underway. The Earthbound cosmologies of water and land, steel and concrete, fire and ice. I want a fungal cosmology. I'm greedy, I want it to already exist. But the new worlds within this one already have counternarratives at their thresholds. Their building has been underway since that palm heel hit that first African drum in North America, spores scattering from the skin with each fateful slap.

Are our lineages mycelial rather than arboreal? Am I a fruiting body resulting from Spores in those sails rather than a leaf on a branch of a tree? It stands to reason, since it's the mycelium that does the true root work of a tree, family or otherwise.

In fingering my way through the wild tangle of this country's roots to find the poison that dealt death and the remedies that made life, I find rich soils in my homes: in Jamaica, Florida, and Massachusetts. But I have to dig for them. At fifteen, in tenth grade, I would have been enrolled in social studies at my high school. I would have learned about ancient civilizations, the Middle Ages and the Renaissance, the Middle Passage, the battle for America, the Trail of Tears and industrialization, give or take a few other "pivotal" world events. In this telling, the processions of my dead, my kin, are mostly mute: their cowardice, valor, messinesses, beauty, despicability, honor, love, treaties, wars, genius . . . circumscribed to a footnote in a European, sugar-money-sponsored story.

My dead, and the other unlucky, would mostly have been mentioned in the moment just after their former glory: their conquest or enslavement, or the theft of their land. Meanwhile, they reanimate their creaky, pale, tired dead for us, the same great men stiffly delivering orders to kill from powdered, perfumed rooms on film and in books and in the same archives reproduced over and over again across continents and decades. I am not singing this creaky, tired tune because I think it's news to you; I just mean to say that I do not wonder when kids with my story want to die.

The next year I would get Chinua Achebe as a bandage on a mortal wound. Where else can we go to find ourselves again except in the interstices and peripheries of the official story? How can we find the ancestral power that imparts dignity when our education is sponsored by the new sugar? Who will walk barefoot in the bush?

The most valuable education I've ever received was a multivalent recounting of the Haitian Revolution. This college class, New World Revolutions, provided the first set of answers to questions my mind had been silently screaming since the state began "teaching" me—what did it look like when we died on our feet rather than living on our knees? The best therapy I've ever received was another retelling, Creating the Transnational Caribbean, another class I took in college. Though Howard Zinn's *A People's History of the United States* had begun the process a couple years before, here was the first time I could reanimate my dead. In this class I would learn that the dead are never silent, they just want listening to. *Their* dead get reanimated just to commit the same sins forever, consigning us to footnotes and silences, our skulls whole coral reefs. But mine are unstoried and free. My dream is like Alejo Carpentier's dream, in which herbalists, the keepers of medicines and poisons, are consultants to those enslaved who, backed into corners and chained, said, finally, blessedly, "Naw."[7] There were those enslaved who made it so that big men could "no longer get up and go."[8] Unstuck in time, I loose the jaws of my ancestors hanging from their gibbets so that they say words that have been attributed to Emiliano Zapata:

"Prefiero morir de pie que vivir de rodillas"
"Mepɛ sɛ miwu wɔ me nan so sen sɛ mɛtra me nkotodwe so"
"Mwen pito mouri sou pye m pase viv sou jenou m"
"Mi prefer dead pon mi foot dem than to live on mi knees"

These words and the deeds that accompany them, the fact that they could have been spoken from any periphery, from outside the frame of any official story, was instrumental in saving my life.

There was a woman they called Mad Mary who used to frequent the road that passed by my grandmother's house. Back in those days a steady stream of people greeted my uncle and sometimes Mamma as they went by under the warming sun of the morning or the hot sun of the afternoon or the mellowing sun of the evening:

"How you do Ms. Darlin'?"

"Consi? Whe g'wan?"

But when Mary went by, what one heard was staccato glossolalia, like an interruption of the regularly scheduled programming, as she headed to or away from town. I could rarely make out individual phrases in her unending stream of words without apparent object, but sometimes they would open up to include commentary on whatever was happening on the street as she moved past—maybe a group of kids using a cricket bat to roll a wheel down the street would find their actions briefly narrated. "You favor Mad Mary" was a common rebuke in our household, a way to chasten the one described so that they quit acting insane.

When I met Mad Mary on the road she always paused what she was saying to look into my eyes or look through me. When I thought she was looking at me, I felt proud somehow. Her plaits ran off in every direction and were always neatly combed. I do not remember her clothing looking anything but clean and well-arrayed. I think of her now while newly prescribed pharmaceuticals move through my veins, providing a buffer between me and the brink, or between me and Mary. Maybe it's the memory of her eyes, still so soft after whatever horror, maybe it's that my name is a variation of hers that causes me not to believe in that buffer, not really.

Picture this: you are an enslaved woman laboring on a sugar plantation. You sometimes practice what they'll later call *petit marronage*—small bursts of freedom in the bush or in the city, skulking on a neighboring plantation to tryst with a lover, hiding in plain sight at the marketplace in town where you hawk the food your master has allowed you to grow for his own budgetary reasons. When you are out among other Black Jamaicans, you are indistinguishable from the approved traveler. There are other women who do this too, you recognize them, they give you foraging tips picked up from elsewhere on the island, from the other petit maroons who haunt the seashore and get news from the next parish over. They say when your people were first brought here, they were chosen for their ability to work the land, but working it is not the same as knowing it intimately, knowing where the wild things grow. The road is rough. It is harrowing. A couple of times there were scrapes that found you crawling from a ditch, looking for a place to wash the junjo from your face. You always pay for this dearly upon your return, but deem the trouble worth it for the exploration, the pleasure, the bit of money the ground provisions you sold were able to get you.

You don't think of these excursions as rebellions per se, but if the voice from the future, from abroad, that you can sometimes hear when you're near the little white duppy umbrellas that make a meaty broth when boiled, asked you, you would probably tell her that it was, since it's become clear that's who she needs you to be. You think of them more like an exchange between you and Master, a game he'll let you play until he won't. The game he'll always win. And what happens when he's finally had enough and wishes to dispose of you? Better to have lived before he killed you!

Your cousin, who works in the Great House in the master's kitchen, is aghast at your behavior and takes every opportunity to tell you so, but she does not complain when you bring her cloth or fresh herbs from town. You do not think this is the life you were born to live, but anyway thank Providence that you were born the type of person who, unlike your cousin, knows you don't need permission.

Reader, imagine you are her, with all you know of the wickedness of those who rule over you. Would you tell your master the few little things you know about the land and how to survive on it? The living that it is possible to make from the fruits of the fertile soil when one is on the lam? Would you tell your daughter to be forthcoming when surveyed about this same subject by a people who would burn down the very Earth if it meant nobody would try to take it from them? Or would you teach your children to teach their children to keep quiet—if you a piss inna di bush and you see dem a come, siddung inna it—especially if your knowledge included which fruits of the earth could poison?

Most mornings, I walk down a street named for a family whose patriarchs are remembered to history as promoters of "religious freedom," whose business enterprises across the Caribbean enslaved thousands.[9] William Vassall was a cofounder of the Massachusetts Bay Company.[10] "The Massachusetts Bay Company" is a phrasal poison whose nine syllables we had to memorize for a test. Their referent signified and signifies money, the kind that does philanthropy here but manages meat grinders that spit out sugar abroad, our education sponsored by this money, the money mutated by generations of inbreeding, those nine syllables unmoored from the living ground and the bodies it ground up or bent to breaking, laundered to a bone, sugar, white in American memory.

Robin was enslaved by Henry Vassall, William Vassall's son.[11] Mark sought Robin's counsel on the matter of escaping slavery, since Robin had tried and failed to secure his own freedom in a well-publicized theft case that, outcome notwithstanding, proved that Robin had gotten close. The two men live on in history with first names only, like Beyoncé, Rihanna, Madonna. The two men met at night in Boston's North End and planned for Robin to secure arsenic from his master's apothecary (in small doses it can be used to treat syphilis) so that Mark could poison his own master. Three quarters of a century later, the neighborhood where their rendezvous took place would be engulfed in a 25-to-40-foot flood of hot molasses when 2.3 million gallons of it burst from a tank, the sweet dark syrup deadly in its production as in its conception.

Mark and the enslaved woman named Phillis who served their master his food colluded to slip the man arsenic until he sickened and died. Mark and Phillis were found out and sentenced to death. They preferred to die on their feet than to live on their knees. In 1755, Phillis was burned, and Mark hanged, then hung up in a gibbet for all to see. In a letter to the Massachusetts Historical Society, Paul Revere uses the execution spot, where Mark still hung (he'd been coated in tar, so that years later, onlookers remarked that his skin looked unbroken) as a locational reference point along the route of his midnight ride, twenty years later. "After I passed Charlestown Neck and got nearly opposite to where Mark was hung in chains . . ."[12]

Pulsing like a necrotic organ where American freedom is about to be born in 1775 is Mark's gibbeted body. I've picked black trumpets, "trumpets of death," near where Mark's body hung.

In the shadow of the silver crucifixes that moved up and down the roads, green poison, yellow poison, or poison that had no color went creeping along, coming down the kitchen chimneys, slipping through the cracks of locked doors, like some irrepressible creeper seeking shade to turn bodies to shades.

—Alejo Carpentier, *The Kingdom of This World*

I dream of Minetta, a fifteen-year-old enslaved girl who reportedly laughed as she ascended the gallows to be executed for poisoning her "master."[13] The narrator of her legal case said she watched him writhe in agony as the effects of the poison set in with a hardened look on her face. She was apparently unmoved as she was sentenced. I dream that the Obeah men who are said to have supplied women with the means to poison their "masters" might have concocted their potions with local mushrooms. I dream that we never lose our contiguity with the land, even when they outlaw the traditions born of it.

Mushrooms are thought to have evolved poison to deter would-be mycophagists from eating them before they had the opportunity to disperse their spores. Poison, in this case, being a mode of survival. Encountering the deadliest *Amanita* varieties in the woods, whether they damn near glow white in their somnolent grace or provoke with their dun-gray and sickly-sweet scent, is a pleasure. And a reminder that not everything is for us.

Not everything is for us.

Not everything is for us.

IN SEARCH OF

Psilocybe Species

(THE MAGIC MUSHROOM)

On learning my own mind,
my chats with a CEO and
a junjo farmer, wild medicine
and Jamaica as "Wild West,"
the uses of veterans,
illegal nature

They don't belong to anyone. Because they're from God.

—Trinidad Garcia Alvarado, Mazatec curandero, in *The Battle over Psychedelic Therapy's Future*

It's late in the evening and B, my not-yet-ex, is about to leave the house to meet with friends and argue over newly released craft beers or whatever. The October day was warm enough to leave open the door that leads to our screened-in porch, despite the hour, and though the air is missing a customary early-fall chill, leaf rot sweetly perfumes the air. There's even a hint of the fruitily fungal there, if you do the olfactory equivalent of squinting. A kiss goodbye and B is on his way; he'll be gone for several hours. A "don't wait up" situation. Now's my opportunity. I grab the cluster of deli-mustard-yellow mushrooms from where I'd sat them on the thick porch carpet. They've left a brown spore residue behind—I'll learn over the course of handling this species often that they stain fabric. I push the front door but don't quite close it.

The conglomeration of them is impressive. It weighs four pounds, and when I found them they looked like a miniature, aerial view of the mess of helium balloons rich girls used to carry around my high school on their birthdays. There are a hundred individuals, easy. I flip them over and check, for the dozenth time, for the blue-green staining happening where it was cut from the ground. Yep, still there: this is really happening. I've been on the forums every day since I found them, cross-checked my identification multiple times, private-messaged an expert or seven, read several discussion threads on the species from across the internet, and I feel like I'm ready to see what these creatures are made of. These creatures commonly called "laughing gyms."

I take a handful of mushrooms, split each in half, and put them in boiling water. They bob like rafts on the stormy, two-quart, stainless steel sea. Watching them change the color of the water to a subtle honey amber, I can't help but laugh at the thought that if I'd made a soup out of the whole lot of them, as I was advised to do by someone who confidently called them honey mushrooms earlier in the week, and had somehow managed to mask their reportedly bitter taste, my night could have quickly become an ER staff's "guess what the chowderheads out in Metrowest are eating now" story. I've made a playlist of wordless, soothing piano music, summonable with the click of a button so there's no fumbling through the danger zone of the internet should I urgently need a distraction. (Are my favorite songs from *The Very Best of Enya* the only exceptions to the lyricless rule? Who can say?) I've put eyes on where exactly the fatty Greek yogurt is in the fridge, if my body does its usual intoxicated thing of telling me, like someone's filter-less aunt, that I could do with a meal. The lights are dimmed to a perfect level. The living room is its usual, Cape-Cod-cottage cave-ish temperature. When the tea has steeped long enough, I am ready.

Ask a group of internet mycophiles what the defining characteristic of this mushroom is and they will say, in unison, "bitter!" This is why many a commentator never got a sufficient dosage to go to the trouble with this genus, which remains obscure despite the experience I was about to have. But bitter don't scare me, no sir. I was raised on tea made from cerasee, a plant also known as bitter melon, a Jamaican answer to every conceivable ailment, and was not about to be cowed by having a bad taste in my mouth for a minute. The gym tea is as earthy as it is bitter. I've established elsewhere that I'm a partisan of dirt's culinary merits, so this does not represent a hardship for me.

Before the action starts, I'm on the porch again, ripping the rest of the cluster apart. Imagine you're eviscerating a stress ball—this sound is nearly identical to that and much more satisfying. It begins to rain. Cluster individuated, I grab a small electric fan from the upstairs guest bedroom and train it on the mushrooms, which I've spread on old tea towels laid out on the floor. It's been days since I harvested them in a suburban edgeland between a trail and a distant neighbor's very grand home, and I want to make sure another hungry fungus doesn't opportunistically move in on my score like any scavenger would. If they're not dried by morning, I'll pop them in the oven at the lowest temperature it will go. I shut the door and lock it, lay on my back on a faux-fur blanket in the middle of the living room, and listen to my heart's metronomic petition. I close my eyes.

Many of the mushroom hunters with whom I first found community, especially those of a certain vintage, got into the hobby as young people looking for the free party growing from wood chips in the strip mall parking island, or from cow shit on farmland whose "no trespassing" signs they ignored. They stuck around in the hobby because other types of mushrooms, then other organisms, then whole ecosystems proved fascinating on further inspection, well after the party ended. A "gateway drug," if that gateway is a threshold to the whole damn world and a practice of awe. And how could you not be awestruck by such a heady first encounter, by that thrill ride starting up, and not knowing what the compounds will show you on that tour of your own mind or of the re-enchanted world at large?

B, the ex, works in the tech industry and has for decades. When I think of the continuum from traditional use to outright abuse of this medicine, on one end are those indigenous to North America and Africa for whom the ritualistic consumption of mushrooms is sacred, and on the other are tech bros he works with who look to the medicine as a shortcut to innovation.[1] When B and I watched the satirical *Silicon Valley* together, he had a hard time with the show, mostly scrolling his phone as I laughed. "These aren't jokes," he groused. "This is just how it is." Snapchat CEO Evan Spiegel agreed, suggesting that the show is "basically a documentary."[2]

In one episode, Erlich Bachman, a man for whom bong rips are hourly deep-breathing exercises, hounds his partner in the startup at the heart of the show to get him to embark on a "vision quest" in order to find a new name for the company.[3] When main character Richard refuses to "eat a bunch of drugs and sit out in the desert and hope some name randomly pops into my head," Bachman says, "Well then I question your leadership." Bachman, who looks something like a fleshy Jesus, takes matters into his own hands, eating a dry handful of "golden caps" (really, Bachman? No peanut butter or citrus?) that had been secreted in a Cherry Garcia pint in the incubator fridge, heading out to the desert with the intention of doing what Richard wouldn't. Only Bachman doesn't make it to his desired destination, not bodily. He's stuck in a traffic jam moments after ingesting his heroic dose. Nevertheless we're in his mind with him as he sits in lotus position on the "desert floor" on the come-up, agave-like plants snaking deliriously behind him, sprouting the logos of extant tech companies as he blurts things like "technolojesus," then fixates on the word "cloud," after which things quickly go sideways. (An aside: I remember when "cloud" was freshly appropriated by the tech world, when the idea of verbally confusing a natural element with a brand still carried the faint, quaint odor of recognizable dystopia.)

Certain sectors of the mycological community would take one look at Bachman and call him a "wook." The wook label attaches itself to the bum, the insufficiently scientific, the person prone to magical thinking. It's here in this murk, where a white man's long hair spills over his Baja jacket as he extols the virtues of plant medicine despite the lack of double-blind, peer-reviewed studies supporting his claims, that this person becomes an avatar for a world that makes a particular kind of Western mind profoundly uncomfortable. A world he hasn't inherited but that he gleefully appropriates. Where the metaphysical is real.

Before I took my first dose of gyms, I read every word of the paltry information that was available to me on the internet and written in a language I could understand. The forums were awash in speculation, and those looking to get high had seen this or heard that somewhere before, but the relationships between these secondhand anecdotes, opinions, and hard data weren't always clear. I spent a while trying to sort out scientific debates about lookalike species and North American vs. European varieties. It is very possible that a psychoactive mushroom I've consumed was only recently described by Western science.

It was all quite a tangle, but it was also a multivalent education that sprang from wanting to know what I was putting into my body. As a friend put it, "Even the fairly well-studied mushrooms are kind of poorly understood, when it comes to the compounds they do produce or that they *can* produce." But if I were to wait for detailed chemical studies of these fringe mushrooms to materialize, I would first see what's left of my youth dwindle down to nothing. I was reassured by an account of the oldheads on the forums giggling through a mycological conference together, someone having found a cluster of gyms that they generously shared with the group before a particularly dry lecture. I identified with those who recounted their experience of the vast variability in the potency and effects of gyms—from the mild visual echoing, glimmering, and distortions that come with a smile that becomes one's resting expression, to the almost sub-perceptual lightening of whatever crises seemed imminent, to the slowing down of one's attention and prioritization of beauty, of connectedness, of "I'm alight." I enjoyed the gourmet recipes that poisons expert Spike shared during a bumper season for *Gymnopilus*: peanut butter and *Gymnopilus subspectabilis* cups, Laughing Lemon Bars, Laughing Gym Gingerbread Ice Cream Sandwiches, and more.

Gymnopilus is a mycophile's drug or medicine, complicated and misunderstood even by those who study fungi. Note that I would *not* recommend others play at the bleeding edge of research, which is where our knowledge of these mushrooms' chemistry is at. I'm also not going to misrepresent my experience with gyms: at times it did feel like I was giving myself the guinea pig treatment. But I was a lab rat in good company, with people to whom I've regularly (and successfully) entrusted my life in the form of mushroom ID support.

Even if my use of gyms is an ongoing hypothesis, an educated guess, they work on me. For days after I've taken a small dose, I am more even-keeled and able to focus. In the woods, which is where I have elected to drink this medicine, I'm attuned to the call of the hermit thrush (perhaps my favorite sound in the world) and likelier to follow my instinct to recline on a rock and sun myself like a lizard, remembering that life's all right.

The compounds I use to cautiously experiment on myself are elsewhere becoming big business—including in Jamaica, where psilocybin-containing mushrooms are not illegal, and thus there is money to be made. MycoMeditations is one of the psilocybin-assisted therapy centers popping up on my native island like . . . well, you know. After I decide to visit the center and see for myself, I check out the website, which includes a helpful framework for figuring out whether to retreat in Jamaica or Oregon.

The page cites the unambiguous legal status of the mushrooms in Jamaica as one point in the country's favor. The fact that there are no limits to dosing in Jamaica, whereas Oregon limits doses to a paltry five grams is also, surprise, a point in this center's favor. On the flip side, this voicey list of pros and cons admits to potential customers that "for many, Oregon comes with the comfort and familiarity of home, while a tropical destination like Jamaica offers adventure and the time to shift away from the regular patterns of life." As I read, I wonder who exactly the intended audience for this website is—who finds Oregon "homey" and Jamaica "adventurous"? This hypothetical customer appears in sharper relief when the website proceeds: "Retreat locations are far from the crime in urban centers, which is the root cause of the [travel] advisories. There is minimal time spent in the larger cities beyond a quick airport pick up." I'm not retreating today, but if I were, a double occupancy room and a week's stay, along with before- and after-care, would cost me US$5,950, while the single occupancy option would set me back $9,200.

When I travel to MycoMeditations on Treasure Beach, I come with plenty of questions about how the outfit will benefit Jamaicans, but also with a deep desire to be convinced it will.

The minute I arrive at the retreat center, I'm asked how I feel about pelicans.

"I love pelicans?" is what I say out loud while inwardly interrogating whether I have ever consciously thought about pelicans.

(The mental report-back says that no, that was a heron in Miyazaki's *The Boy and the Heron*, and that oh, yes, actually—whenever that video of one trying to swallow a duck resurfaces, I do root for that video's star, even as I deeply don't. Pelican't. Much to Pelican's chagrin.)

My reflexive response gets me led into an office space where company CEO Justin Townsend and Director of Operations Abbie Townsend have been looking after a sick young bird since the day before, a bird who looks smaller and stranger when not gliding over the seashore. Needing only to hear "you can pe . . ." I am instantly squatting, hearing about the bird's probable dehydration and youth, dismissing an apology for the way it smells strongly of digested fish in the office (apparently not how it always smells), enjoying the softness of the feathers on the bird's head against my fingertips, having my welling eyes captured in a photograph Justin asks if he can take. Allowing me to pet a (non-drugged) wild animal as part of introductions has, I must admit, set an impossible standard for all subsequent attempts to start out on the right foot with me in this life.

Justin is Myco's second CEO since the company's modest beginnings round about 2014, when it hosted its first retreats on a local family's property. Since then, it has become "one of the longest-serving psychedelic-assisted retreat centers in the world," per their website.[4] There are retreats underway at the time of my visit, which means that I can't tour the site proper, though said website and documentaries on YouTube show that it looks like a typical luxury resort. It is oceanfront, of course, and from the palm-roofed cabana where Justin and I have our conversation, between the offices and the beach, I can look out at the sparkling aquamarine expanse of the ocean. My brothers and I drove here through chilly green bamboo and red clay mountain morning, but by 10 a.m. the seaside heat is already baking the oils on my skin. (My brothers have been reluctantly banished to the beach.)

Justin is long of beard, bespectacled, with close-cropped hair, an ink-heavy tattoo on his arm and the sand-kohled toes of a man who spends the day in sandals. He found psychedelics as a depressed and anxious teen growing up in a small village in Southeast England, using them recreationally, freaking out, but realizing that despite the freakouts, some of the pressure from his depression and anxiety had lessened after each session. A familiar story about a kid wanting to explore the contours of his own mind. He studied civil engineering at a small school in Oxford, then spent a couple decades working with startups and in venture capital to help "promising, innovative" companies figure out how to maximize their profits. Still questing for ways to manage his mental health in work environments designed to destroy it, he made private studies of Jungian depth psychology (which observes the deep psyche or unconscious, exploring the wellspring of our imaginations and the stories we tell ourselves as humans in an attempt to deploy those in healing—a radical lens on the world) and transpersonal psychology. He spent two years learning holotropic breathwork, which allows the practitioner to reach altered states of consciousness with controlled breathing. He later taught breathwork in Germany on a retreat model.

The story of his move from rat race to professional healer is well-formed. It contains shades of "I'm not only the Hair Club For Men president, I'm also a client." He came to Myco for a retreat a few years ago, took over from a fellow who was perhaps not as well-suited to running the sort of business Myco was poised to become, then brought all his training and experience with investors to bear on rebuilding the ship such that Gwyneth Paltrow might want to visit it someday—which she did in 2020 for her Netflix show. MycoMeditations represents the longest sustained stint on Justin's résumé. That the other stints were short makes sense for someone who has made a career of advising companies on how to be the most profitable versions of themselves before moving on. In the company but not of it. His LinkedIn lists New York City as his location.

Justin took Myco's reins from former CEO Eric Osborne, who used to run a gourmet mushroom farm in Indiana while disseminating psilocybin underground, a fact which landed him in jail for a week and lost him that farm.[5] He made his way to Jamaica after that and administered retreats from the home of a local Treasure Beach family whose patriarch is a Rasta, which is the only way I can make sense of the way Eric, who is white, speaks to those on the town's neighborhood forum. There was a thread for locals concerned about the retreat's arrival in 2015. Rather in the style of Ras Trent (Andy Samberg's hilarious sendup of white rastas), Eric accuses the naysayers who wish he would keep his junjo nonsense in Negril of "chanting down progress." Channeling Marcus Garvey, Osborne assures them that "Jamaica itself is the Black Star Liner. . . . Fear not . . . I have bigger plans outside of yard."[6] Eric currently runs the Psanctuary in Louisville, Kentucky, an organization with the tax-exempt status of a church, from whose website he offers the following descriptions of his debacle: "While in the conceptual stages of establishing the Church of Soma in 2015, without religious protection, I was arrested for providing access to the sacrament." That sacrament, of course, is psilocybin mushrooms. So you leave the place where your practice has made you a criminal, start a company where your practice is legal, dose hundreds of people and garner a reputation, sell the business to a VC, and come back home in a blaze of glory to start your own church? Got it.

When Justin Townsend and I start talking about the business's relation to the local community, he is able to point to education work MycoMeditations is doing in local schools, employment of locals, partnerships across the Caribbean, drives for Christmas gifts for neighborhood kids, as well as funds for sanitary products and swim lessons. According to Myco's website, their operation infused US$1 million into the Jamaican economy in 2023.

But for me, a highlight of the conversation is when Justin speaks of a fisherman who fell into a treatment-resistant depression when COVID hit and tourists weren't coming in. We'll call him D. The lockdown put D in dire straits, since the bulk of his business came from selling seafood to restaurants. (Treasure Beach is a fishing village whose industry is not currently experiencing a boom, to say the least.) COVID pushed D to the brink, and it sounded like he would have tried anything, even duppy umbrella, to pull back from it. Word got out to the community from staff members at the retreat center who witnessed patrons coming in "looking like the walking dead" but leaving "alive again." So, this fisherman had some indication that whatever went on at Myco's Blue Marlin location was doing *something*. Nevertheless, he was a strict Seventh Day Adventist, probably not in the habit of fixing problems with what the UN considered Schedule I substances. So he put Justin through the wringer of speaking with his wife and adult kids about the ins and outs of psychedelic therapy. Once they got the family seal of approval, Justin worked with D's doctor to get him off the ineffective antidepressants. By then, enough trust had been built to send him on his trip, which Myco supplied for free. Despite the cost to Myco, Justin says that when D is "fishing again, he might come by with a couple of lobster, or a fish," into which I interject my enthusiasm for all things barter, before he continues, "I would actually take his lobster or take his fish because it gives him his dignity, you know?"

D's outlook and affect did a 180. While his circumstances hadn't changed drastically, his perspective on them shifted to let in some extra-normal light. That's one effect the mushrooms can have: switching the camera angle and the lighting, imparting vantages that may have been hard to come by otherwise, thereby laying the groundwork for emotional and spiritual healing. The term "entheogen" was coined by a group of ethnobotanists in 1979 in order to capture this effect of being infused with the divine ("begetting God in" the consumer).[7] The perspective gained can work like God in the mind, while seeming to come from outside of it. Justin's recollection of D's transformation has me shockingly near tears, but I pretend to type on my laptop there in the cabana, looking away to cover it up. It's not just that I'm sleep deprived and thinking of that sick pelican—what took place for D is what I want for any Jamaican who wants it. But even if it was in Myco's business model to supply one life-changing trip to a local for every ten paying customers, Justin says that there isn't a whole lot of demand for their services locally, and that Myco is "in the community but not of it."

Nevertheless, he tells me, there are psychologists based in Kingston who sent about three or four of their final-year students to Myco to train in the use of psychedelics, which Justin said they are now administering in their practice. Slowly, slowly, the spores disseminate. Other mental health professionals have gotten wind of this "new" treatment and are referring their patients to those original three or four. Justin says that it's good it's happening at all, obviously, but that he's impatient with the molasses-slow pace. In that moment it's easy to believe he really is impatient, that he's interested in the struggle to bring the medicine to Jamaicans. I wonder how long it could take to shift attitudes about mushrooms among Jamaicans, which puts me in mind of Valentina and Gordon Wasson's 1957 article in *Life* magazine about their trip to Oaxaca and the sacred mushroom rituals they observed among Mazatec people there. How it set off the first "psychedelic revolution" in the United States, and what it took for Americans to be open to these new possibilities, before that potential was abruptly foreclosed at the start of the war on drugs.

Myco seems to be trying to be a good citizen of its community. In terms of clients sharing their experiences afterward, there could scarcely be a more sterling track record than their reviews on Tripadvisor, which remark how their protocol lines up with (emerging) industry best practices. In my book, it's certainly ahead of another luxury therapy center in Jamaica which shall remain nameless but inhabits, I shit you not, a *former goddamned plantation*. MycoMeditations seems to be the gold standard for a legal, luxury retreat with a medical modality. So why am I salty?

When I ask how Justin contends with his own colonial heritage, about the potential for what he's doing to be read as a neocolonial endeavor, his answer does not quite map onto my question. I've asked about what it means for a white man to come to a place where the per capita GDP is around US$10,000 in order to extract the "natural resource" of their permissive drug laws.[8] He answers by citing his founding of an association meant to regulate the industry and educate Jamaicans, and his joining of a government body tasked with similar goals. Perhaps I should have made it clearer that my highest hopes for the place where I was born have nothing to do with the imposition of rules according to the interests of (mostly foreign) capital.

The bottom line is that there's a lot of money being made from the healing they're doing on Treasure Beach. And as with anything where investors are the final word, and profits are the biggest measure of success, educational efforts and whether the poor want the service or not will remain dramatically beside the point. The other, secret bottom line is that when Justin first came to Jamaica, he tells me, it was "the Wild West"—there was no psychedelics industry to speak of. Interestingly, he does not say there were no psychedelic sales, and mentions in passing Miss Brown's Tea House of Negril, a place where backpackers (yeah, okay) and locals alike (!) would get mushroom tea back in the '60s. As he passed on to another topic, my mind highlighted this in flashing neon.

Later, I do a little digging and find that Miss Brown's (sometimes called Mrs. Brown's) is now Tedd's Shroom Boom (alternatively spelled Ted's on the internet, though the sign that marks the physical spot across from a Texaco has two *d*'s). When I call the establishment, a groggy-sounding man answers—he admits that I've awoken him but is game to talk. He says Tedd's has been in business for "fifty-odd years." They offer takeaway and delivery options, and will even help you with dosing if you ask. The home page of their website has a photograph of a man with a silver goatee, his Hawaiian-style Incredible Hulk shirt open to the navel, more than just the suggestion of a six-pack in evidence. (Does a man need hair on top of his head to qualify as a silver fox?!) He's wearing one of those downward smirks like when you're trying not to smile, and is flanked by four broadly cheesing young women, also Black, whose enthusiasm suggests they're the ones who asked for the picture, under a sign that reads: "Tedd's Shroom Boom: Mushroom Tea, Ice Cold Beers, Mushroom Punch." Next to the words, a gray hand holds a cluster of graceful fungal fruiting bodies, underneath which is the mysterious yet decisive label of "Original."

Tedd's isn't the only game in town. Negril is known as a beachside psilocybin mecca throughout Jamaica (as evidenced by the gripers in the Treasure Beach forum) and to backpackers around the world who don't mind the informality and lack of quality assurance of a transaction with a rando on the beach. Among the conservative populace, that comes with all the negative associations of the "mad hippie" tourists or "expats" playing it fast and loose with decency and their own mental well-being on beaches they've ruined with their presence.

There are a number of spots where you can purchase dried *Psilocybe* species, get it delivered to your hotel or home, drink prepared mushroom tea (à la Tedd's), or buy a day pass for an outpatient type of session with oversight. I am delighted to discover that Mr. and Mrs. Shrooms, also known as Shawn and Shelley Ann, run an establishment that offers the latter for as low as $200, according to their website.

"We actually have a history here—it's part of the culture of Negril," Shelley Ann told me when we spoke. This was the first time I had ever heard it framed as if that scene were a history one would want to point to, and I perk up—it sounds like a reclamation, a flipping of the usual narrative. That usual narrative, of course, is that "dem a come from foreign an' a mash up fi wi beach dem," but the more I think about it, the more that seems like a mistake. In all likelihood, fringe Jamaicans—artists, religious minorities, and the like—were right there with the "mad" white hippies, but were perhaps not seen as the primary customers. For Mr. and Mrs. Shrooms, it's a shrewd way to conceive of those sixty years, those decades leading up to this moment Justin dismissed as the "Wild West." Because of course for that analogy to work, one has to kill off or ghettoize those who were already there.

Shelley says that the Jamaicans they serve at Mr. and Mrs. Shrooms have to be a little more hush-hush about all the good the shrooms have done them, due to the stigma. But she hopes that will change once there's enough consolidated research available, and members of the government see the promise being realized at places like Myco and her own establishment. "Unfortunately, the negative stories are the ones that get highlighted," she says.

Even though the medicine is challenging, Shelley Ann believes Jamaicans are equal to the task, as long as they go into the experience with their eyes and hearts open, having done the required reading, and proceed with a willingness to integrate what they've uncovered into their lives beyond the sessions themselves. "If yu want good, yu nose haffi run," she says of that cost. Their service consists of a screening call, a tour of their grow facility, tailor-made "set and setting" for your two- to three-hour session with a facilitator, and supplies for microdosing.

Shelley Ann told me that she and her husband used to work for one of the bigger retreat games in town, but didn't love how that outfit conducted business, especially how the mushroom farmers were treated. They looked into what it would take to acquire licensure themselves, realized it was within reach, and went for it. During our chat it became clear that Shelley's own history with both larger doses and microdosing makes this story personal for her, as well. "I'm no therapist, but I have plenty of experience in hospitality," she says, and I imagine all the healers and bush doctors and curanderas throughout the ages, the ways their practices have been systematically devalued in order to make way for Western modalities that ultimately end up stealing from those traditions they denigrate, with nary a nod to the source. I think of the *sabia* María Sabina herself, dying poor and malnourished in Huatla de Jiménez after the visitation by Valentina and Robert Gordon Wasson that is credited with bringing psilocybin to the West.[9]

What will it mean to legalize, and then medicalize, psilocybin? Will facilities boasting highly paid healers (doctors, nurses, psychologists, and psychiatrists) be seen as having a monopoly on using "science" in their practice? Non-Western healers have been using psilocybin as medicine successfully for millennia, of course, without such credentials, and their shamanic models, which Justin suggested were less well-suited to a Western mind, don't tend to be used at the luxury spas. But do you have to go to school to study what goes on in the mind of a tripping person to anticipate their need? Can't you train rigorously "on the job" to minister to the most challenging trips—a fact that Justin Townsend himself, who trained as a civil engineer but now facilitates psilocybin sessions, is living proof of?

When I found out about Mr. and Mrs. Shroom, I was pretty quickly on the phone to my older brother to suggest we pay them a visit next time we're back home together. We both have things we'd like to work through, and I knew the prospect of a local establishment in Jamaica would appeal to him. The setting, at least, would be very different to the urban park where his last trip took place, where his ego died and the cowboy boots of the person one bathroom stall over positively *became* Pennywise's clown shoes. "I didn't like that trip, man," he told me. He contrasted it to other, gentler rides where he felt more in control of himself and at peace in the environment, in one case looking through a telescope at the rings of Saturn and feeling pleasantly and exceedingly small. The trip in the park where he didn't feel safe, though?

"I was a passenger in my own mind. A public place is not where I want to feel like that." I would have to agree—to be in the midst of feeling content in one's insignificance, as a Black man, and then to suddenly hear sirens? It's enough to stop a heart, or at least put unseemly pressure on the bowels.

The classic setting for recreational or personal medicinal mushroom use is deep in the woods, but this is not a location everyone can access or feel comfortable in. In the article "Race as a Component of Set and Setting: How Experiences of Race Can Influence Psychedelic Experience," Logan Neitzke-Spruill uses research from social psychology and sociology of medicine to call attention to the ways race relations and racialized self-perceptions affect set and setting. This arena is, of course, comparatively and criminally understudied.[10]

Something Shelley Ann said about Jamaicans being "behind" the rest of the world when it comes to the adoption of "new" technologies (evidenced, according to her, by culinary mushrooms very slowly appearing in supermarkets as junjo wins mainstream acceptance on the island), got me thinking about what it means to call Jamaica "the West." Who does it, and to what aspects of the culture do they refer when they do so? What if adopting the concept of Westernness (always already fraught) and seeking out its spoils is a matter of survival . . . yet that impulse to bring a culturally syncretic place into complete conformity with "Western" ideals will almost always, thankfully, fail?

First of all, what is this slippery, ill-defined West? It certainly doesn't derive its meaning from geographical considerations. It could mean "exemplary of whiteness." In "Toward a Caribbean Psychology: An African-Centered Approach," psychologist and professor Marcia Elizabeth Sutherland summarizes Western psychology by pointing to values like universalism, individualism, rationalism, mind-body dualism, the experimental method, and so on.[11] Sutherland's concern is that these cultural norms do not fit neatly with Caribbean culture or psychology, springing as it does more from Africa than Europe. She adds, "Western psychologists have long-standing contemptuous views of people of African descent." Anecdotally, just about every Black person I know has been failed by mental health professionals, who have missed spectacularly obvious diagnoses, thrown the wrong pills at misdiagnosed ailments, and doubted the self-reports of individuals living with their own symptoms.

To be clear, Jamaicanness looks like a form of Westernness the way some people live it, and, for many others, does not—at least not all the time. For every hellscape of paperwork a Jamaican must fill out to receive the simplest government service, or every child passing their A levels, there is an Anansi story being told, or a farmer practicing agroforestry and tending herbs for bush medicine. For better or worse, however, the Jamaican government can be relied upon to import systems that work for monied visitors and not Jamaicans themselves.

One factor Sutherland uses to distinguish Jamaican psychology, which she says derives from African social ideals, from its Western counterpart is collectivism and extended family. Think D making Justin meet his family. She says that a possible explanation for contemporary Caribbean violence, for instance, is the breakdown of the kinship networks that used to govern social life. There are several such realities of Jamaican thoughtways that might not match up with Western models of "the" psyche, and any regulation on psilocybin that seeks to codify engagement with the medicine on Jamaican soil without taking Jamaican psychology into consideration will be just another colonial experiment doomed to failure. Remember when we talked about junjo and how the traditions of a people can be overwritten by colonizing forces? How there were certainly Jamaicans eating culinary mushrooms, but the Brits' own mycophobia likely cast long shadows over that practice? Don't let anyone tell you that someone who came along in 2017 was the first to take advantage of Jamaica's psilocybin laws. It was simply a cottage industry before the luxury retreat model arrived.

After Justin leaves me in the cabana to wait for the lead mushroom cultivator to arrive, I pull up the *Psychedelics Today* blog I found that morning when I Googled "Taíno" and "MycoMeditations" together. I'd searched these terms after noticing Myco's website said their Blue Marlin location was built on "a registered archaeological site with a rich history of the Taino people," and I'm glad I did. Apparently, a former trip facilitator here reported that retreaters sometimes saw "Taíno people" in their visions without necessarily knowing who they were. The tribe's elongated skulls, the result of cranial shaping, were a clear identity marker, but the meaning of their visitations during multiple trips was left indeterminate.[12] I'm reminded of another story I found, about a Taíno cacique, or chief, on Quisqueya (later Hispaniola) who was said to have allowed Columbus to sit in the *duho* (a seat of power) that *he* sat in when he traveled to the spirit world after ritually inhaling psychedelic snuff called *kohoba*.[13] This has been haunting me all morning, and I'm dazed and goosebumped by the thought when Sidi arrives. The young woman was supposed to give me a tour, and Justin had also encouraged her to share her personal connection to mushrooms with me.

Sidi Genus, who has a pretty dope junjo farmer name, was born and raised in Treasure Beach. That family whose property Eric Osborne first ran the fledgling MycoMeditations from? She is its daughter. When she was fifteen, Sidi was awarded a soccer scholarship in the States. Her parents were against her accepting it, thinking she should finish out high school before touring the world. But they were overruled by their willful daughter.

America was not what television had led her to believe (girl, same), and her host family did little to show her around the place she'd been so eager to explore. According to Sidi, they were not nice people. Her life with them consisted of school, soccer practice, and home. She regretted her decision as she sank into a deep depression—suffering from anxiety as well, having regular panic attacks.

"I was the most unhappy person in the world!" she exclaimed mid-interview. By the time she was matched with other hosts, more agreeable people on the whole, the color had already been drained from the world, and the ruts of an unfulfilling routine were dug deep: school, soccer practice, home. In both Washington, DC, where she finished out high school, and Mobile, Alabama, where she got another scholarship to go to college, those were the only settings and habits of mind: school, soccer practice, home, all sapped of meaning. At the beginning of college, she studied nursing; by the end it was sociology and criminology. A refrain of Sidi's throughout this telling is "nothing made sense." In light of this she made the decision to go home to Jamaica, where her friends and family were surprised to see her back.

Sidi gives us a tour of the staff offices. (My brothers, now back, tag along for this part—I know they've noticed our tour guide's beauty and guess they're also wondering what's in the goody bag Justin has left me with, given the mycelium surrounding us.) There are three or four other Jamaicans milling around as we nose about the orderly growing facilities with spawn bags full of Guinea grass and horse manure, a transparent tent with a flow hood in it at the corner of a room, a couple of large autoclaves, and some "inventory" in a fancy stacking dehydrator in the corner.

Sidi is very comfortable in the lab setting with all of its surgical precision, owing perhaps to her nurse's training, but it's easy to see she thinks of mushroom growing as an art, too. We head out to the greenhouse, a large black mesh walk-in tent with bags of spawn hanging from strings tied to metal rungs. The space is glorious, and while the most recent batch has mostly been picked, there are some stragglers at the very back that we get to examine up close. The strain of *Psilocybe cubensis* we're looking at is called B+. "Be Positive." Get it?

When, as a young woman, Sidi moved back to Treasure Beach, MycoMeditations was a much humbler outfit, in the process of starting up. Sidi was skeptical of the whole mushroom thing—she associated them with the culinary and the hippie and didn't see how there could possibly be therapeutic applications for any of that. The only other mushroom, as far as she was concerned, was the puffball that grew wild in the sandy soil, which would turn gangrenous black and release its spore cache, known locally as "duppy powder." In Jamaica, fungi are often identified with "ghosts, obeah, witchcraft, all that stuff," so any disentangling of mushrooms from that litany of prejudices was going to be painstaking work. How to go from that to an understanding of their healing power? But Sidi watched as the patrons of Myco were transformed during their brief stays, heard them tell their stories. Still, it remained largely academic to her.

When she was offered a job, she took it, not out of any deep conviction but for the same reason people have taken jobs since the invention of money. After around a month of working with the mushrooms, she decided to try them. Right on the heels of this decision, however, was the realization that she was pregnant, which forestalled the first dose. When she was finally able to take it, she transformed into her brother—who had drowned at the beach near her home when she was seven or eight—*in the very midst of his drowning*. She says she experienced all the emotions of a violently dying child. It was difficult, to say the least. But she emerged from the recesses of her own subconscious at peace. The medicine had allowed her to cast off the guilt of having survived her brother, guilt she hadn't even known she felt. After a single dose.

That's not all this single dose did, either. As a new mom with terrible anxiety, Sidi says the paranoia was near debilitating—every corner held a new danger for her baby, so intensely that it didn't feel normal. Explaining how this was tempered after that first dose, Sidi speaks of the fungi the way so many mushroom people around the world do, according them agency: she says they "told" her, simply, to calm down. It was like a switch had been flipped, expanding her capacity to be a present parent, able to let go of some of her old catastrophizing narratives that pitted everyday life against her son. "I thought I was going to have to quit my job, because if I wasn't looking at my baby, he was going to die . . ." But a single dose changed that.

It's prudent to be wary of the thing that sounds like a magic bullet. But to hear Sidi say that in performing the hardest job on the planet, of nurturing and being changed by a new human, she was able to temper her anxiety from a nine on her personal scale to a five, with a single dose—it's something else. I want folks like Sidi and Shelley Ann to be the future of mushroom love in Jamaica.

For her part, Sidi received an intimation of her life's work as the medicine did its own work to heal her. The message she received in the healing, that "everyone needs mushrooms," now helps animate her life. To that end, Sidi has partnered with Darren Le Baron, a Barbadian psychedelic researcher and educator, to bring mushroom cultivation workshops to Treasure Beach. As I write, the two professionals are preparing to host the second coming of Shroomshop Jamaica Retreat, a six-day workshop where attendees learn to grow psychedelic, medicinal, and culinary mushrooms, take nature walks complete with foraging and mushroom identification, and learn about mushrooms in ritual ceremony. On the signup page for the event there are local scholarships available for these experiences, as well as a link people can use to sponsor a local to attend.

Later, at home, my mind remains abuzz with all of these conversations. In the time since my visit, the complexities and simplicities of monetizing the sacred, of neocolonial inheritances, of who gets to benefit from the fruits of their own land (even when the fruits are legal concepts) have turned to a stew that sits in my stomach, impossible to digest. I'm struck anew by a point Justin made about how to build bipartisan support for psilocybin therapy in the States: how on the Right it's all war all the time, but what of the veterans who come back home, broken, to no support? He believes that the best way to bridge the divide between conservatives and liberals regarding legalization is through efforts to heal veterans, the police, and first responders. I'm recalling this now because it reminds me of one specific veteran I met six months prior.

When I met Teresa Schwinghamer at Telluride Mushroom Festival, the largest gathering of myco-nerds, mushroom chefs, and psychonauts in North America, I made a beeline for her friendship. I felt my awkwardness dissipate in her company, relaxing in a way I seldom can with new people. I was both in my element (among the mushroom-obsessed) and outside it (in a stunning ski village and playground for the rich in Colorado), but her presence helped me feel at home. She's effusive and warm, with a sudden laugh you want to prompt, and knows just how to style a patchwork kimono. We coled a foray together, fussed about getting to events on time, and did debriefs on how each day went. Still, it took a couple of days before I learned how mushrooms saved Teresa's life.

Teresa's upbringing in the Philippines and later experience in the Air Force, where one of her tasks was performing drug tests in a medical lab, reinforced the idea that "drugs" of any variety were for "losers," those who were "lazy, dangerous, and vagrant." It took a while to unlearn those lessons. But a turning point came in 2014, during an intense depressive episode, when her mind got set on sinking. From a morass of suicidal ideation, in a monumental show of strength, Teresa used psilocybin with the intention of "sitting with herself" for the first time. She had tried psychedelics before, but mostly as an escape. For this trip, however, the stakes were much higher. Sink or swim.

She had what many would call a bad trip, but the medicine let her examine wounds that hadn't healed correctly: injuries from the unchecked bipolar disorder that runs in her family, alongside a familial belief that mental health issues are make-believe. She describes the process as purging: "not the throwing up kind, but the tears . . . purging emotions by feeling through them." Finally, she was able to assess the thing in her mind that was necrotic as separate from the healthy tissue of the rest of her life. She went from sleeping all day and crying whenever she was awake to wanting to fight. Soon after, she enacted a tradition of taking psilocybin on New Year's Eve, and then conceived her goal of becoming a therapist who uses plant medicine in her practice. The only thing in her way was a felony conviction for a crime she didn't commit.

I'll let her explain in her own words, which she did at a Colorado Springs City Council meeting in 2022, giving public comment on why her state should have "access to natural medicine."[14]

In the video of the meeting, Teresa says, "My felony is the result of a year I spent running a Cannabis Church with my husband. He and I thought our establishment was legally operated, and we did our best to provide the community with a safe spot to procure and use cannabis. The man who invested with us to start the Cannabis Church was also involved with several other businesses, but was apparently also starting to sell large quantities of cocaine and other drugs, something we weren't aware of at the time. Our investor's activities resulted in his house and businesses, including ours, being raided by the police. My husband and I were arrested eventually, as was the investor. We cooperated as fully as possible. After many years, the end result was this: The investor trafficking cocaine was able to extend his court dates perpetually, and ultimately all charges were dropped. My husband, a Caucasian, and the registered owner, was told his felony would be dropped after completing two years of probation. I, the woman of color that worked there, was told my charges would not be dropped, I would have three years of probation, and a bigger fine.

"I tell this to you not for sympathy, but to impress upon you an important point. Many people who have felonies have them because of America's failed 'war on drugs.' As my story highlights, the people most likely to receive these felonies are not the hardened criminal masterminds you seek to keep out of this new industry. Rather, it is the disenfranchised and those unable to advocate for themselves that routinely fall victim to a system that all too readily victimizes them because they are easier targets."

Teresa shared her story, hard as it was, because she wanted to support an equitable approach to decriminalization of psychedelics in Colorado. The bottom line for her is that the roll-out of psilocybin must be done the right way, so that the substances are brought to the market responsibly. She wants access to be equitable. She does not want the capitalists to swoop in with money from out of state, to steamroll over the underground market for mushrooms as they did on the cannabis market in Colorado, though she fears the worst.

She had favored a bill (Initiative 61) that would have focused more narrowly on the decriminalization process rather than a sweeping legalization, protecting those who are already cultivators and practitioners, and thus benefiting less privileged people to a greater degree. Some say such people aren't ready to handle the responsibility of growing the industry, even though they have managed a robust black market—sometimes ending up in jail for their trouble. The legislation that ultimately passed in Colorado was Proposition 122, advocated by New Approach, a lobbying group from outside the state.[15] This well-funded PAC is seen by some as a front for corporate interests poised to leverage their money to dominate the psilocybin industry in Colorado in much the same way outside interests took over the cannabis industry.

When Teresa first told me she was in school to become a therapist who administers plant medicine, I felt that satisfaction you feel when someone who is perfect for the labor pursues it. She's been in that place she'll need to help pull others from. It's like Justin Townsend asked: "Would you rather have some overcompassionate therapist who's twenty-eight years old . . . or do you want someone to walk with you who's been to hell and back and had to claw their own way out?"

As it stands at the time of this writing, though, Colorado law wouldn't allow her to use psychedelic therapy in her therapeutic practice—because she's a felon. Knowing this, it's hard not to think of Eric Osborne, the first CEO of MycoMeditations and now the head of a church that doles out an illegal substance in a state where the laws aren't even ambiguous. Despite the loss of one legislative battle, Teresa is still in school, and plans to simultaneously fight for her right to administer the medicine that saved her.

Here's what I remember best about my first experience with Gymnopilus, alone in my then-suburban home. First, there are visual effects. The living room is vivacious with life—the artwork, depicting a sea vessel attacked by a monster sea creature, becomes animate. This is not an often-reported effect of gym ingestion, and I panic a little, worried what is to come will be too much. I pour myself some Black Maple Hill bourbon to get familiar sensations into the mix. I put on the Enya and that won't do. *The Best of Django Reinhardt*, no. Mingus's *Ah Um*, definitely not. I deviate from the wordless playlist and try "Because" from the Beatles *Love* album and that also won't do. I try "Dreamland" by Marcia Griffiths and anticipate the delusion in "and surely, we'll never die," before she even sings it, experiencing those words as ominous, rather than sweetly foolish, for the first time. Okay, music is out. Against the odds, though, the bourbon works to mellow out the experience a little. I lay down on the carpet and listen to my heartbeat merge with the buzz of all the electronics in the house. Breathe in, breathe out. I feel my face relax into a smile. I want to look at it in the mirror but know that's a bad idea. I ride out the impulse. I continue to mellow. Various thoughts I might have had, whose meanings I may or may not have held lightly: The president is tweeting about how there was "no collusion," I still think hourly about the woman in Illinois who carved out a hollow in my belly to live in long-distance then broke off contact, I am wasting my life.

Maybe an hour in, I would love to take a walk, but I turn on the television instead and here's where things get strange. The visual effects have slowed to a stop so the only way to explain this is that it's actually happening—Comedy Central is showing the episode of *Broad City* where Ilana and Abbi take mushrooms and go on their walkabout through New York City. I turn the television off. I turn it back on again. They're still there. I switch it off, finally, before the pet death I know is coming. The rest of the evening is spent sketching, inscrutable charcoal figures that I still have hidden away in a closet somewhere. I rock back and forth, openly stimming, still smiling. I'm wasting my life but that's not possible because I'm here breathing just like everything else, and what I actually mean, perhaps, is that I want to need and be needed by a community beyond these walls. I go into the yard with a flashlight and collect twigs and leaves to arrange into little scenes that I'll take apart and reassemble again. I feel the lightest I've felt in months.

During the first COVID lockdown, when everyone else was gleefully tucking into their problem drinking by pretending mixology was their new passion, I quit mine, abrupt as a runner's heart attack. Simply stopped enjoying alcohol's effects, though I'd been known within my circles as the consummate drinking buddy: shutting down bars with restaurant industry friends as if the tap were bottomless, racking up partners I had met at a bar, airing midnight grievances with whiskey's borrowed courage. But trapped in my apartment during those early days of the pandemic with three fairly reclusive dudes and a scratch-happy black cat named Panther for housemates, I lost the taste for it. The glass of wine that usually went with dinner, and would be followed by a couple more as I wound down from the day's ghostwriting gig, was now a vile desiccant, leaching all my life-sustaining water and replacing it with dust.

In this apartment, I was not cohabiting with a partner for the first time in my adult life. I biked to the woods every other day with *Gymnopilus* tea made from stock I'd foraged the year before. This, in my inexpert opinion, is the reason I quit drinking then—the small doses of mushrooms were treating the anxiety that stemmed from my then-undiagnosed ADHD, the anxiety that alcohol had helped me escape. In this new, four-walls-only alternate reality, without the need to mask for a long-term partner who didn't understand ("mask" as in "pretend my brain worked in the expected ways"), I was free to slow down and heal. I'd drink my tea and wander for hours, dazed by the fine grain of deeper noticing, the usually rigid outlines of lichen-crusted boulders vibrating with what felt like my own pulse-beat. My mind was plastic as I drank oak- and pine-leaf-filtered air deep into my lungs—especially heady in a world afraid to breathe.

These jaunts constituted something more than microdosing but certainly something less than tripping—because I could feel the effects of the chemicals moving through me and they worked on my perception in appreciable ways, I was technically intoxicated each time. But I noticed that after each of these walks, for several days I felt equanimity. From the time I was a teen until I finally said to hell with it in my early twenties, I was put on every antidepressant known to science for my occasionally paralytic behavior and nervous breakdowns. Nothing worked. The immediate, dramatic, and symbolically beautiful results that these *Gymnopilus* were having on my mood and outlook for several days at a time remade my world and altered what the word "home" means for me. How could I not feel at home in the place that showed me the flesh of the earth, fruiting abundantly in the least likely places, which in the process of living its own earthly aims gave me the right and the responsibility to mend my mind? On days when I find medicine blithely growing from the ground, I feel the land's intention toward me—its volition.

For the past few years, my partner and I have driven to Northampton, Massachusetts, on my early February birthday, to eat good food and hike nearby trails. Coincidentally, as of the first year of this tradition, Northampton was one of around ten US cities to decriminalize psilocybin. I wasn't consciously thinking about that vote as I brewed the gym tea in our hotel. We were primed to explore, as this was our first excursion in our new car, the Blubaru, since the start of COVID. I only wanted to start this new trip around the sun off with a pliable brain that was ready to receive the year with grace. We drank the tea and walked the Robert Frost Trail in the Connecticut River Valley. Robert Frost, that great noticer of New England woodland beauty whose "miles to go before I sleep" often comes, unbidden, to mind when I'm woods-walking after sunset and the owls have started to wonder about me. My partner loves and will often recite lines of Frost's "Birches" while on the trail named for him:

> It's when I'm weary of considerations,
>
> And life is too much like a pathless wood
>
> Where your face burns and tickles with the cobwebs
>
> Broken across it, and one eye is weeping
>
> From a twig's having lashed across it open.
>
> I'd like to get away from earth awhile
>
> And then come back to it and begin over.[16]

IN SEARCH OF

Fomes excavatus

(THE TINDER FUNGUS)

On visible fire, bearing an ancient torch, the fungi who eat coal, mushroom reparations, collective rage

Fire has acquired the vigor, subtlety, and endless variety of the organic world.

—Stephen Pyne

I don't remember who said it or who the person who said it was talking to or whether to say it was appropriate in context or if the person who said it was imaginary, as in hailing from a novel or TV show, but I was once very struck when someone remarked, "Thank God fire burns visibly."

I must've been old enough that my mind could turn to science to make sense of those words, extrapolating what concepts underlay them even if my grasp of those concepts was incomplete—*light spectrum, the human eye, how fire has been beneficial to us throughout history*—but I must also have been young enough that the sentiment haunted me like an echo your brain decides to keep, inspiring awe, leaving something chewy and indigestible, a scorched room in the limbic edifice, to come back to over and over again over the years. *Thank God fire burns visibly.* How large the yet-inarticulable can loom in the mind of a child—a mind which struggles to ask: Wait, is the universe random? Does fire also burn invisibly? "Thank God"—is this a prayer of gratitude to our laws of physics—are you supposed to be grateful for those? Is gravity pissed I don't regularly thank it? Do coincidences happen on planet Earth? What is the nature of visibility? Did we just happen to get a fire that we could see? What would have happened if firelight never decided to show up to the part of the spectrum that we could make out? This place where thanking God makes sense—this portion of a broader spectrum—this skin of our teeth—does it suggest the existence of an entire, incalculable shadow world inside our own where other processes are shaping life as we know it, but our ability to perceive these processes can come down to mere luck? To a brief flicker someone was there to see, an efflorescence standing as proof of the chemical churn and gurgle governing everything?

The very relativity of it all might have been enough, but taken together with who was being invoked at the center of that relativity—fire itself—I was never going to be able to look away from *Thank God fire burns visibly*. Fire, that bringer of punishment and warmth, that maker of phoenixes and hearths, that righteous cleanser, that act of war, that sign of the apocalypse, that evolutionary powerhouse, that deity.

Two lines from songs that often get stuck in my head are "hot like fire / take you higher" from Aaliyah's song and "it's too old and cold and settled in its ways here" in Joni's.

When my mother considers the reasons she wanted to move to the States to give her children a better life, she thinks, quite specifically, of the hard work it took to bring fire into the kitchen uppa Mahoney every day.

"And now all your daughter wants to do is go to the woods and build fires, huh?" I snark. She continues as though I haven't said a thing—perhaps because it's rarely a surprise when we kids do the opposite of what she set us up to do. When she was coming up, country kitchens were separate structures. The walls of my grandmother's kitchen were made of wattle, that is, panels of woven bamboo that could breathe out heat and smoke but keep out other elements. You could just peek through the slatting, but not really discern what you were looking at, nor could someone look in from the street to see what you were up to. The stove structure itself was a bamboo table with clay daub molded to it. Large stones, then eventually cinderblocks, were used as firebreaks. Metal sheeting was put over these stones. This setup was rebuilt a few times during my mother's childhood. The fire, of course, had to be fed continuously in order to keep the food cooking at an even temperature. Stocking the wood and kindling was a tedious chore—finding branches and logs dry enough to keep the fire going was hard, especially in the wet of a place where each morning one comes face to face with clouds. My mother says that in order to avoid performing all that labor in the dark, dinner was finished early, sometime in late afternoon. This aspect of their formative years makes sense of an odd bit of teasing from my aunt: when we were on a WhatsApp call and I told her I was making dinner at 8:00 p.m., she said I was "making bug food."

By the time my mother was in nursing school and living in the city, the family had graduated to a kerosene stove with two burners like a hot plate with legs, and a coal kiln with a grate on which to place pots. But burning wood always remained an option in that kitchen, when coal or other fuel was scarce. During my visits back home every summer, I distinctly remember my grandmother's clothes redolent of woodsmoke, her grimace as she called me into that crucible to bring her something, the smoke reddening her eyes.

My family's early memory of fire is also one of trauma. My mother's father, drunk and lumbering; a kerosene lamp that served to light their one-room home; the room in flames; her mother's hands the extinguisher. My grandma's hands burned and bandaged, at work soon after, painfully weaving baskets to sell at market so the children could eat. Fire (the feeding of it, its terrible extinguishing) may quite literally be the element that destined me to a hyphenate identity—the reason for my mother's migration. My mother, on a plane, looking out on Boston's leafless winter trees, thinking there must have been a terrible fire. But she had come to this place where fire stayed banked and also at the ready, at the turn of a knob or the push of a button.

What happens in your mind when you imagine a fire, out of doors, not contained by a pit or stone oven, actively burning and spreading and smoking, hungrily lapping at every piece of fuel in its path? Chances are, if you are in and of the Christian God–haunted West, you might panic. Maybe try to remember everything you learned about fire safety—"stop, drop, and roll," that "only you can prevent" the ones in forests, to "never throw water on a grease" one, etc. But what if that fire was late to their job? What if they're the only one on the goddamn plane who can deliver the baby, and you're standing in the aisle, hindering their progress, talkin' 'bout "For behold, the Lord will come with fire, and with his chariots like a whirlwind, to render his anger with fury, and his rebuke with flames of fire"?[1] As a Westerner, you may not know of the trees in your area that need fire to trigger seed release, of those among your neighbors, the deer and bears, who depend on the biodiversity engendered by flames for food and habitat. You might associate it only with destruction, with the ache in your lungs from so-called megafires. With the "This is fine" dog from the comic strip, whose home is engulfed in flames as he sits there idly with his cup of coffee. With the carbonized disk in your oven that was once a frozen pizza. With the photos of kangaroos, in blazing hellscapes, who look like they want to slap you and, honestly, they have every right. With the charred frame of the police car standing as proof of the chemical churn and gurgle governing everything. But fire, like most else, is not binary.

Disintegration. Cleansing. Coevolution with trees, plants, wood as specialty. This all sounds familiar. Fungi eat charcoal. They are among the first signs of life after a forest fire has burned and sterilized everything else.[2] Oyster mushrooms are used to convert toxins left in the wake of a blaze into usable nutrients for other life.[3] Fungal fruit bodies have been used to carry fire from place to place by hominids for millennia.[4] Black truffle mycelia chemically burn the earth around the tree they've colonized as a not so subtle "fuck you" to vegetation that would crowd them out.[5] Some activists burning police motorcycles, earth movers, cement trucks, and the like, to keep Cop City from being built, have fungi as their mascot—as we'll see. Melanin makes fungal tissue fire tolerant.[6] Fire, as I said, is not binary. It is destructive, constructive, the thing that marked us as human, and, as we change the climate to make it hotter, drier, a thing that may mark our end.

Laetiporus sulphureus blazing from a log, orange and yellow, a common beacon in my home woods; *Flammulina velutipes*, literally "little flame with velvet feet," periscoping from an elm snag, antifreeze in its tissue on a chill day in November, a fungus that has been to space;[7] *Lactarius pyrogalus*, literally "fire milk," whose flesh can scorch tongues, feeding and being fed by a hornbeam. What work they do—fire, fungi—what glorious work in common!

To make whole worlds smolder in memory, to move and lick and lap in shadow, to fruit sudden and vivid upon alchemical request. To take oxygen, to expel carbon dioxide. Fungi and fire are kin, which means, I guess, that I am family to fire.

Combustion can be seen as the ancestor to breathing.[8] Fungi, by helping plants onto land, oxygenated the atmosphere, and animals learned to breathe that poison, their respiration a slow burn.

Homo sapiens and those like us have wielded fire for round about a million years. Some folks say that you can date the start of the Anthropocene to this moment, the sticks igniting in those early people's hands—the Promethean dance of the blaze in their eyes—the world changed in an instant, slowly. Some say the real shift came later and could more profitably be called the Plantationocene, but that word is hideous and most people aren't going to want to say it.[9] I wasn't there so I'm not doing the Hollywood thing where we assume early hominids were dull because we have cell phones and they didn't, but it's sort of funny to imagine those first fires, to have watched them from a distance, how much trial and error might have gone into their mastering, how many heads were cocked to the side and scratched as the accidentally large conflagration started toward camp. But fire wasn't a beginning. Likewise, this particular point in time cannot have been a destination. If I crane my neck back far enough to see a future part of this loop we're trapped in, the invention of fire is there again.

Ötzi the Iceman knew fire, how to summon it. Ötzi, our best preserved European human mummy, discovered in the Alps in 1991, 5,000 years after his death, perhaps even knew how to carry it from place to place as a coal in his backpack, because he also knew some mycology. It's wild to me that in so many descriptions of what Ötzi had near him when he died, writers have left out the two varieties of mushroom he toted with such care. His tinder fungi, in which fire can nest during the journey from campsite to campsite, and his medicinal birch polypores, both just as mummified as he.[10] As far as I'm concerned, these are what lend homie the most gravitas. The accounts that baffle me usually make much of the bow and arrows he was in the process of making, his copper axe, his shoes insulated with grass, stuff like that. Fine, okay, the injuries to his frame do bespeak an intensely violent last few days. He probably bled out from the arrow wounds to his back, after all. And it's neat that Time took good care of his Chalcolithic UGGs. You would listen to that true crime podcast—maybe you already have. But the mushrooms! I hope that when future archaeologists find my plaid backpack still clinging to my skeleton, they are underwhelmed by my utility knife and enthralled by my polypores.

We are such a stark minority in all of human history, those of us whose lives are divorced from the blazing that has characterized life on earth for millennia. I think about all the invisible fire that haunts our days—heating systems for houses, various cooking implements in our kitchens, irons, blow dryers and hot combs (I shudder at the memory of leaving the hot comb caddy plugged in on our side porch when I was a teenager, of the potential for burning down all my mother had built, of the fact that she actually did burn the hell out of her hand, leaving a handprint of burnt flesh behind on the caddy), our lamps, our boilers, the internal combustion of our vehicles . . . the list goes on. Our firearms. To live a modern life is to have everything mediated, all that hot work invisibilized.

In some ways, we're like bored kids with matches. Eighty to ninety percent of the wildfires that rage today in the US are started by humans. Some of those are accidental, some are arson; these numbers tend to spike around the Fourth of July.

When I was a bored kid, one image that transfixed me was the iconic moment from *Waiting to Exhale*: Angela Bassett walking away from a fire she has set as vengeance against her cheating husband. The movie, you may not remember, is set in Phoenix, Arizona, which in 1995 had not yet arrived at a year-long fire season, which is to say, no season. Now every day is another chance for flame to catch.

The scene was a living mirror image of the spore of the rage that would reside in me the rest of my life. A Black rage. One difference, though, is that the sorts of feelings I regularly feel that make me want to destroy another's property are generally not provoked by scorned love. Which is not to say I'm above that, just that my heat-of-the-moment sabotage is likelier to be a mind game or seven, while the things that make me want to set a fire are communal.

A high-intensity wildfire (it doesn't matter which one) was hot enough to vaporize most aboveground life in its domain. (Its digestive haze eventually blew across the country to give me a headache, to turn my New England sky sickly orange-pink well before sunset.) The fire had moved slowly through the chaparral, destructive and leisurely as a hurricane, burning deep, leaving a foot of ash behind. Maybe we lost a species of insect we didn't even know we had. Maybe we lost a hundred, given the thousands "discovered" as new to Western science every year. There's no food or habitat for some of the surviving vertebrates—the coyotes, the horned toads—to come back to. But come back they will. Soonish. Those mice who could burrow deep enough underground emerged into the light to see a landscape made of brittle, spent coals. Most life seems carbonized, turned black and weightless, wanting to fall away from itself on the wind as ash. Most life. But cue those fearless, sleepless, deathless tricksters.

The fire had barely breathed its last breath before a survivor, the single-celled yeast *Geminibasidium*, was in full bloom, in all its stunning heat tolerance.[11] It blooms and blooms and dies quickly, after which (and after rain) comes *Pyronema*, an ascomycete that perhaps survives due to its hard-as-nails, melanated sclerotia (is it obnoxious to wink at you whenever melanin does hard work?), quickly becoming a dominant player in the soil. *P.*'s mycelial filaments move through the wreckage, eating away at the exploded, fire-roasted necromass (plant matter, dead bacteria, dead fungi) left behind, creating a webbed mat, fruiting what looks like lumpy orange fuzzed skin . . . but place your nose right up against it to see it's made up of individual cups or cushions. It clumps the soil beneath the black, keeping it moist, making it harder for the individual grains of pyrolyzed matter to fly off and make a life somewhere else. Of course soil bacteria are at play here, too, and in finding their own footing, help to remake their home's equilibrium. *Pyronema* remains the dominant fungus for around three months, blooming, eating, matting.

Charcoal is a tough sell for any organism to metabolize. We are right to marvel at *P.*'s apparent willingness to do so. If *P.* is eating what most other organisms can't, when most other organisms can't compete and would only be caught dead in that whole mess—if *P.* is truly munching on charcoal, then its role in biotic recovery after fire is key.[12] *Pyronema*'s forbearance of flames, its quickness to colonize and willingness to work with material that other organisms find distasteful suit it to the work of bringing diverse life back to the soil after an inferno. It's as if these humble, vibrant recyclers were made to prepare the ashes before those phoenixes of flora and fauna can even begin to think about rising.

How do we know all this (or fruitfully hypothesize) about *Pyronema* and its kin? Part of our gratitude should go to the human mess of the Rim Fire of 2013, started by a hunter's illegal campfire and contained after nine weeks, a "natural" experiment that gave ecologists the datasets to compare pre- and post-fire mycobiota.[13] Another portion of the thanks belongs to those observing soil activity after prescribed burns. Still another share of props goes to the burn nerds constructing pyrocosms: that is, labs "filled with forest soil and wired to thermocouples" and then set alight, simulating fire's effects on said soil.[14] Yes, scientists are really out here creating miniature forest fires in buckets. Pyrocosms look like what my little brother and I might have gotten up to in our backyard as kids if there had been a toy with a thermometer in it that we could have dissected (maybe a doll that got warm with fever?—the '90s were weird) to extract the tool for our own purposes. It sometimes seems like scientists, same as artists, are just overgrown kids who managed to spin their backyard experiments into a way of life. When I picture blazing buckets used to predict the future of species succession on recently sterilized land, it looks like the realm of witches, of strange alchemy. But these witches publish their spells and undergo peer review. What a time to be alive.

Upon learning about the pyrocosm and its spirit of righteous play, my mind immediately journeyed to Atlanta, to a forest the Muscogee called Weelaunee (later dubbed the South River Forest, then *re*-dubbed Weelaunee) where, as of this writing, the construction of the infamous "Cop City" has been delayed by two years. Cop City is the nickname of an imagined "Public Safety Training Center," its own kind of pyrocosm, with a realistic planned replica of an Atlanta neighborhood, shooting range, burn building, area for bomb testing, and place to land Black Hawk helicopters. Everything the police can dream of to further militarize their operations, right next to a lower income Black neighborhood. Masterminded by a private organization called the Atlanta Police Foundation (APF), Cop City is a place for the Atlanta Police Department to get better at terrorizing the Black people who still outnumber other populations in that municipality, despite a steady decline due to gentrification and other factors. Black folks also make up the majority of the opponents to the scheme, according to an Emory University poll, though plenty of other demographics have also spoken up and put their bodies on the line to oppose it. At one early meeting, seventeen hours of public comment were heard on the center—people were opposed to it by a 2:1 ratio.[15]

Many, many people don't want to see the $90 million police (and fire) training facility built on eighty-five acres of land that already perform important ecological functions as forest (cooling urban temperatures, absorbing potential floodwaters: jobs becoming ever more important as the climate changes). So a decentralized coalition of Georgia residents and outside agitators alike, from police abolitionists to forest defenders, have funneled their voices through megaphones, sat their bodies in trees, set fire to earth movers, locked themselves to cranes, staged concerts, served food, held skillshares and Shabbat ceremonies and witches' bonfires in the woods, collected signatures, and generally refused to back down from their position that "Cop City will never be built." The vigorous community opposition to this project has not relented since the APF publicly moved toward construction, hot on the heels of the 2020 uprisings ignited by George Floyd's murder. *Thank God fire burns visibly.*

This ongoing movement, sometimes called Stop Cop City, sometimes Defend the Atlanta Forest, was the first in a long time to make me want to quit my job and step into the path of the machinery. I've been inspired by how both its underground and visible apparatuses adapt, how the nodes of its network are free to act autonomously for the protection of human and more-than-human freedom, how it dramatizes the inextricability of those freedoms. It's thrilling to watch resistance blooming and to see folks fired up—here, finally, a serious critical mass and international attention.

So many eyes are trained on the machine's methodology as it burns our future, leaving the ashes of Black, brown, and other poor bodies in its wake. The machine only knows how to make hellfire—it does not trade in the cleansing sort. It both runs on and generates the toxic necromass of white supremacy. It knows punishment, not regeneration. So how can we oppose its dystopian vision? How can we know what urban life can look like, post-crucible?

We can look to the free barbecues and puppet shows held in the Weelaunee Forest, and to the mutual aid that rose from the burnt police precinct in Minneapolis. People taking care of each other, unmediated by those who profit from our isolation. Fire and fungi both work to clear what's inert from our systems, what no longer serves, what suffocates our bodies and earth, our bodies, the earth, one and the same.

On a trip to Atlanta in May 2022, I met up with a couple of local activists I'll call V and K, who said they would give me a tour of the Weelaunee Forest.

"We basically want to make the cops afraid of the woods," V told me, as we neared the area where the forest defenders had set up camp. It was a surprising reminder that cops are more vulnerable than we tend to think of them as being, perhaps more fearful, given the carte blanche they apparently have to kill, untouched by the law they purport to defend.

I asked if V or K knew whether the activists ever spared a thought to the fungi working literally all around them. K said that soon after the plans for Cop City were announced and resistance began taking shape, she was talking with other activists about how quickly different groups started organizing marches and other events independently but simultaneously, acting autonomously yet with shared goals. "One person," K continued, "said we're like a mycelial network."

I thought of how Merlin Sheldrake contrasted our animal bodies with mycelium: "If we cut off our head or stop our heart, we're finished. A mycelial network has no head and no brain. Fungi, like plants, are decentralized organisms. There are no operational centers, no capital cities, no seats of government."[16] This is the hardest thing for authorities to understand about actually popular movements like this one: there are no leaders whose removal will stop the organism. *We are the metaphors for them.*

The pine trees towered, some sporting banners and Antifascist Action flags. One said "Forest Defense Is Self Defense," another "No Forest, No Peace #StopCopCity," while still another acknowledged the Muscogee and Cherokee peoples' traditional stewardship of the earth we walked. The mid-May day sweltered around us, temperatures in the nineties. Clearly, what this city needed was some of its trees cut. Because of the slant of the news from Atlanta lately, I was worried the outlook was dim. The camp's quiet, and the sodden expression of some of the few defenders milling about, seemed to corroborate my worries. But K was clear. "We're not going to lose," she said, with finality.

A couple of months before, Georgia's government had charged dozens of forest defenders with domestic terrorism (hypocrites much?) for anything from attending a woodland concert to camping on the Cop City construction site.[17] These activities got placed right alongside the property destruction that might more predictably land one in legal trouble. (The domestic terrorism law under which they were charged, by the by, was passed in reaction to white supremacist Dylann Roof's attack on Mother Emanuel Church in Charleston.)[18] After our visit, in September 2023, the collecting of bail funds and other mutual aid for arrested activists resulted in dozens of activists being arraigned on RICO (Racketeer Influenced and Corrupt Organization Act) charges, originally intended to combat organized crime.[19] This financial intimidation and carceral vengeance is ironic, but it's also a sign of the state's fear that movements to hold it accountable to its constituents will continue to gain traction.

Atlanta's officials aren't even trying to conceal the link between the 2020 uprisings around the country and their outsized response to those pushing back against an expansion in policing. According to the ACLU's website, Atlanta's attorney general's office made its position plain: "To make its case, the indictment relies on people's beliefs and community organizing as the connective tissue for sweeping criminal liability. It devotes 25 pages to vilifying Defend the Atlanta Forest (DTAF), the grassroots movement opposing Cop City's construction, identifying its 'beginnings' in the nationwide protests against George Floyd's murder and protests in Georgia against the police killing of Atlanta resident Rayshard Brooks, and calling out the movement's 'anarchist ideals.' It paints the provision of mutual aid, the advocacy of collectivism, and even the publishing of zines as hallmarks of a criminal enterprise."[20]

The Atlanta 61 are, as of this writing, still on the hook for racketeering. Twelve of them issued a statement that includes this: "Mutual aid and solidarity are intrinsic to how our Earth and its inhabitants survive—in spite of systems of permanent suffocation, we all breathe the same air—a conspiracy with the trees."[21]

After that first visit to the Weelaunee Forest and news of the RICO charges, I was hungry for encouraging updates, so I made sure to attend an info session in my town in September 2022. The presenter had come up from Atlanta on a tour meant to spread the word like wildfire, and they played to a packed room. I say "played" because of the mischievous, interactive way we were called on to participate: throughout the presentation, the audience was encouraged to issue a resounding "boo" whenever the story's villains were mentioned—the Atlanta Police Foundation, Mayor Dickens, the bought-and-paid-for city council. We said a collective "nice" when Z* shared organizers' efforts to bring Weelaunee's neighbors of every stripe into the woods to enjoy the trees and feel their stake in them. We cheered when Z told us how simple it had been to rename Intrenchment Creek Park to Weelaunee People's Forest on Google, a name that reflects its one-time indigenous stewardship and the rematriated future that folks are fighting for. We whooped when we heard how the project had been delayed by targeting contractors and subcontractors with less of a financial stake in the project—sometimes with phone calls, sometimes with judicious applications of fire to heavy equipment.

I spoke with Z after their presentation. They told me that the police department trains for cities with clear lines of sight, built on grids, that they're stuck in boxes born of hierarchy (at this point we'd clearly made a leap away from the merely literal), which meant they could never have any sure footing against the decentralized, autonomous, and multitudinous movement in the shadowy, tangled woods, the one they're trying so hard to quash.

When Z talks about activists engaged in direct action, I get an impression of a scene with which I'm familiar—anarchists who know this is the most important fight of our lives, so they've dropped out of work, school, their families, every system that runs on hierarchies and binaries to instead camp high above the ground in trees and actually exercise the freedom so oft invoked in this country but rarely lived. They lead lives of fire. Watch them touch the earth mover brought to Weelaunee to take down trees and see it catch, the blaze a cultural fire we should be glad so many are still willing to light and maintain. Z was cool, charismatic—we

* Not their real name.

exchanged info because I wanted to hit them up later in the year when I passed through Atlanta on my way to Florida.

On our way down to visit family in South Florida for Christmas, Walter and I did drop off some donated camping supplies for forest defenders, meeting Z at a vegetarian potluck in a community center. (*In a conspiracy with the trees. . . .*) A few weeks later, ready to head back north to Massachusetts, I reached out to Z again and offered to lead a mushroom walk for activists when I passed through Atlanta, and Z spread the word for me. But the day before the walk, January 18, 2023, as my partner and I were still making our way up through the endless flatness of our home state, we received a message from Z asking if we'd heard the news. Folks were unlikely to make it to the walk, planned for a woodland that, in December, one defender had told me "is what Weelaunee could look like in a few decades if left to grow." The reason? An activist camped in the woods that would be bulldozed for Cop City had just been shot and killed by police.

Feeling rage and sadness like twin animals warring in my gut, I accepted my partner's offer to go on to Atlanta and take the walk ourselves, anyway. Photos we took in those wet, unseasonably warm woods the day after the killing show early daffodils and curly onion grass, *Xylobulus frustulatus*, *Stereum ostrea*, a dripping *Entoloma*, and *Pleurotus*, the oyster mushroom, often touted as a remediator of polluted environments.[22] I only learned the nickname of the murdered forest defender later: Tortuguita, "little turtle," a twenty-six-year-old nonbinary person from Venezuela who had dedicated themself to mutual aid and environmental causes.[23] If they had not been shot in their tent fifty-seven times the day before, they might have joined the walk with us.[24] If they had, I wonder which of these species might have caught and held their interest. The last of the pictures I took that day is of a bent "dead end" sign in the middle of the trees.

When I google "Tortuguita Cop City" to see what the official story is around the shooting, the results return with this "People Also Ask" section: "Why was Tortuguita killed? Where was

Tortuguita from? What is Cop City in Atlanta GA? Will Cop City be built?" I answer the first three in my head.

1. Because they grew with a tree, protecting it, when it was marked for destruction.

2. Spare us the outside agitator line—there's no time for it. Anywhere in the world, Tort's brown body was vulnerable to hailing bullets—anywhere in the world, their body was made of the same elements that make up every other living thing. Tortuguita's actions translate to love on every terrain.

3. Cop City is an act of war. In the most generous (to the state) terms, it's a "public safety training facility" paid for with taxpayers' dollars that the majority of Atlanta's residents don't want.

I'm curious to see the search engine's opinion of whether it will be built, so I click. It offers up Wikipedia: Once complete, the center is planned to be used as a training campus for police and fire services and is expected to open at the end of 2023. This means, dear reader, that they are well behind schedule.

Later that month, at a vigil for Tortuguita held in my city, I lit several tea lights and placed them under graffiti that said "Don't call it public safety," and "ACAB." During a collective moment of silence among the dozen of us who assembled, I imagined a future where a little girl who looks a lot like I did at her age has a snow globe in her hands. But instead of snowing snow, it rains fire. And instead of a little town, at its center is a cop car in flames. And once the fire settles to ash, a succession of fungi, then plants, then animals pours from the burnt-out frame, life muscling through a former instrument of death.

Many months later, at a gathering where folks discussed the radical things they planned to do to protect forests and waters and parks and the vulnerable all over the country, a portion of

Tortuguita's ashes were ritually spread over the cool water of a river as comrades from Atlanta sang low, words I couldn't make out. I watched the proceedings from a distance, moved beyond measure, wondering what other bits of earth their body had sanctified.

There is a mushroom in order Cantharellales that the new Audubon guide has christened with the common name of "candlelight vigil."[25] They do indeed look like pale candles with wicks aflame, evenly spaced, growing up through a forest-green layer of algae on fallen wood. The fruit bodies undoubtedly grow from an assemblage of organisms even as they seem separate, even as each one is a bright beacon in the dark.

Life is not predictable when you cobble it together from passing suggestions, niche interests, dreams that shade into delusions, ways of making money under the table, spit, cussedness, and vinegar. That's why I could not have said even a few months before it happened that I would be spending late winter and early spring 2024 on a 700-acre property in Northern Virginia, at the height of spring fire season. During my stay there, at an artist's residency hosted by a garden foundation, my cohort and I were given an ecological tour of a small portion of the foundation's rolling grassland and pastureland, with its patches of old growth and young woods, situated at the southern border of the Loudoun Valley in the Virginia Piedmont. The tour was given over the course of three hours by Rea, the foundation's entomologist, ecology expert, and foremost *Doctor Who* fan.

The day was glorious—it was sunny, one of those spring mornings that tricks you into putting away your winter coat. Though we were all dressed too warmly, the group mood was buoyant, and I had just made Rea laugh by showing her my plushie of an *Ophiocordyceps*-infected ant (an entomologist is probably the second most likely type of person to appreciate such a thing) when there was an interruption. What sounded like several packs of fired up dogs erupted into howling cries nearby. Rea informed us that we were hearing a fox hunt passing through on the other side of the hill behind us. In answer to our widened eyes and cocked heads, she said that the participants were definitely all dressed up for the occasion—I would learn later that this meant tweeds for most of the hunters, a scarlet jacket for the master huntsman, and perhaps velvet hunt caps all around. "That's still done here?" the Brit and German in our cohort wanted to know, and our guide offered a "welcome to America" shrugging apology in the form of the assurance that "they just sort of chase the fox, they don't kill it." I wanted to ask where exactly these Virginians might have picked up that habit—aloud and in the general direction of the Brit—but we'd only known each other a few days.

Once the barking subsided, Rea pointed to the undulating blue horizon. We were flanked on either side by wizened hills, which clearly filled Rea with pride in this place. "They're older than bones."

She said this of the mountains the way you say something you know will take your listener a minute to come fully alive to. It's easy to know intellectually that the Appalachian mountain chain is the oldest in North America, born in violent collisions that began around 480 million years ago.[26] But when your own spine is brought into the conversation, propped up to illustrate the physical boundaries of deep time, and the entity making your humanity feel so impossibly whelplike is right there, you could touch those mountains if you wanted—it somehow hits different. Rea talked about the behavior of the landscape as that of an elder, especially as compared with the histrionics of the mightier mountains out west. After all, those mountains display the bratty, destructive antics of an adolescent—huge landslides, more frequent quakes. The dramatics extend to their wildfires, too: Western ones rage all too frequently across vast expanses of land. That's just not style, the Appalachians might say, admitting that sometimes there'll be a little local three-alarm, but nothing to smoke-signal God about, my gosh, grow up.

Because I had been hiking in nearby Shenandoah National Park for a couple of years now, I associated this region only with the lush forests of oak, hickory, pine, maple, and poplar on those very same hills. But Rea challenged that prejudice by engaging us in a forensic exercise, pointing to the old trees scattered and singular across the landscape.

"What do you notice about these white oaks here, in terms of their branching?" she asked. I perked—while this sort of forensic exercise was natural for an ecological scientist, it had perhaps a too intense appeal to the little girl in me who desperately wanted to be a spy or a Sherlock-style detective. It's just as well life didn't turn out as I'd hoped, since fealty to a government agency would have been tricky for my anarchist heart. Learning the Earth's many languages is a much better use of my busybody tendencies.

I answered that the oaks branch out low, which indicates they were not in competition for sunlight as they were coming up, which means they weren't part of a dense forest, which means that the land we surveyed had not changed too much, in certain respects, from the time the tree was young. This is indeed the consensus about the region, Rea said: that in pre-settlement times, the area we walked was grassland and open savanna. Indeed, before their genocide, the Indigenous people who managed the land did so through agriculture, hunting, and burning.

According to a Southern Forest Resource Assessment report, for centuries annual burning of grasses and less frequent burning of undergrowth kept the land open, with clean lines of sight for hunting.[27] These regimens also prevented shrubs and trees that don't love fire from taking over, thereby encouraging biodiversity. Fire encourages a plurality. Good fire can prevent hellfire.

Here in the West, my eyes have been trained to see the human and the natural as separate and the former as preeminent, so on a cursory glance I would have said the horse pastures and farms surrounding us are this place's most conspicuous physical inheritance. But the foothills laugh. And the Manahoac and Monacan footprint on the place stares back at me, plain as day. I am trying to evolve sharper, more ancient eyes, ones that see like my human and fungal forebears. These would show me that as long as we have been human, we have manipulated the land. Some manipulations have been more sustainable than others. Our history is one of fire, from its judicious use to its suppression. Rea told us that under Native stewardship, and even before it, this place was likely regularly on fire, sparked by lightning and Native maintenance regimens. This made an indelible mark, made the valley a place of plenty, and attracted those who would steal the land.

Later at the residency, transdisciplinary conceptual artist Kieran Myles-Andrés Tverbakk, aka ESO MALFLOR, and I are commiserating about how we narrowly missed the controlled burn the foundation had just done to dissuade invasive plant species in a field on the property. We wished we had been there to watch. "They couldn't have waited?" wonders Kieran. They are at work on a body of pieces called *Controlled Burn*—while here, they carved tree trunks into bones, collected the actual bones of animals, used burning wood and smoke in sketches of trans bodies and landscapes engulfed in the flames of protest, all toward a stunning transformation of their studio space into a temporary installation that had the feel of actually being outside. Kieran's artist's name, MALFLOR, is a reclaimed Mexican slur for queer people (Kieran is nonbinary, of mixed Mexican and Norwegian descent).

One of our first conversations was about the queerness of everything on Earth. They talked about the beauty of the concept of the perfect flower, which has both male and female parts—I, of course, rattled on about the 23,000-plus sexes of *Schizophyllum commune*. The more I learned about Kieran's work, the clearer it became that so many of us are working to dissolve the dividing line between humans and "nature." These residencies are great for a number of reasons, not least because you get other artists thinking about the questions you're asking—the centrality of bones in their pieces meant that I shared with them the coordinates of corpses on the campus, or handed them a bag with a skull and antler shed on their birthday; my preoccupation with fungi meant they texted me photos of crusts and spalting (the intricate black lines indicating a fungal turf war) on the wood they worked. This spalting wove its way through the vertebrae they fashioned from a large log—the fungi enacting their antipathy for one another into inky permanence in the fiber of the wood.

Kieran's work highlights the ways in which our bodies and E/earth (the whole planet and the soil) are one and the same, how this bad-penny separation between "us" and "nature" is at the heart of the way transness is violently denied and othered. When Kieran opens their studio for a final visit from our cohort of painters, performance artists, researchers, composers, and writers, all the sawdust from the bone sculptures has been gathered to mimic soil, with deer and raccoon bones growing from it like plants. On the wall are bodies in transition: the burning landscapes of Minneapolis during the George Floyd uprising—the chest of a friend who has just undergone top surgery. These pieces have quite literally been made using fire. Kieran directed the smoke from a flame they lit on a stick of *palo de ocote* (pinewood) toward the paper. Once the flame was gone, the wood turned to charcoal, which was applied directly to the paper for finer line work. The land and bodyscapes indelibly mirror each other.

Kieran read their artist's statement as we milled about their studio. We were transfixed by what they'd got up to during our five weeks together. Their statement spoke of *palo de ocote*'s use in Mexican cleansing rituals, of the use of natural elements "as a way to draw a simple direct connection between my materials, processes, and subjects." The title of the series, *Controlled Burn*, referred not only to Kieran's process but more generally to an act that might look harmful but was actually "an act of taking control of one's own life or ecosystem in hopes of reducing the potential for a larger catastrophe in the future." This applies to the surgeries that save trans and other lives as well as to selective arson in an uprising. In Minneapolis, they wrote, "the burning down of the third precinct [was] a strategic action intended to restore the natural ecosystem of the neighborhood."

Remember the Target that was infamously looted and burned in 2020, the one close to that ruined precinct? That was Kieran's local Target. They were at the epicenter of the firestorm from the moment it erupted. While Minneapolis burned, and the police abandoned the blazing city, people took to the streets and poured out their palpable rage there. The Target was probably burned because it was a big-box store with likely investments in local policing, but Black- and brown-owned businesses also fell afoul of the conflagration, so that eventually people boarded up their windows with signs that said "immigrant owned" and "children live here" as a reminder. Later, when I asked if Kieran felt afraid as this was happening right outside their door, they said, "The gunshots were a lot. I'm in community with those who were rising up, and we were protecting each other."

Any fear they might have felt was soon subsumed by a practice of care. They began using an award of art grant money for mutual aid—distributing clothing, food, and more to the unhoused and others who needed it in a Sprinter van, dubbed *A Community Arts Bandwagon* (or *ACAB*). In the van traveled an installation composed of Día de los Muertos altars. They also made banners for protests and shields for when the cops decided to reappear to quell the unrest, with their usual tools of clubs, tear gas, and rubber bullets. Once the embers had cooled later that year, the BareBones Puppets Theatre of Minneapolis commissioned artwork from the community as a gesture toward healing. Kieran had already driven through the city and spied rubble they wanted to incorporate into their work, so they were ready, with another artist, to build altars with the remains of local businesses—stacked bricks coated in resin. Stark against a gray November day when it was installed in Minneapolis, the altars sprouted croton, spotted laurel, and poetry in all-caps handwriting. An alebrije snake coiled in repose on a piece of charred wood, guarding. Kieran has seen and crystallized the stages of fire in their work, from spark, to ruin, to rebirth.

About one hundred lightning bolts strike the surface of the earth every second.[28] But most of these moments of contact do not end in fire, just as it won't always be clear what it takes to spark a movement. But while lightning seems spontaneous, it is in fact the result of agitation.

According to *Fire: A Brief History* by Stephen J. Pyne, "the chemistry of respiration is a chemistry of combustion"; the inputs and outputs are the same.[29] Our breath is ancient combustion. That's why they want to snuff it out, or bank it in prisons. Because they know we know.

"I can't breathe" on every poster and T-shirt, chanted by protesters just before the firing of a tear-gas canister. The sheer quantity of breath in the streets sparked the whole country. So they redoubled their efforts toward extinguishing, building bastions for cops. A word often used to describe people taking to the streets is "unrest," which has always been a little funny to me. "Unrest" presupposes "content repose" at times when fire is not burning visibly in the street. But the fuel load is always there. A dangerous fuel load on every street corner, round every cul-de-sac, in every traffic jam, on every high school campus. The gaps between us grow into gulfs and fuel accumulates, piling high. The heat turned up by greed at the top, the magnifying glass of twenty-four-hour news and social media, and poof, up it goes. But what looks like instant conflagration isn't spontaneous. It is never spontaneous.

After the serviceman Aaron Bushnell protested the genocide in Gaza in front of the Israeli embassy in DC by dying in a brief discharge of visible fire, which he ignited himself, I could not stop thinking about it. What is it to set oneself on fire? What is the act? How to incorporate it into my understanding of the moment, such as it is—or do I even try, is it achievable? Who self-immolates, and what has it really meant to the people left behind to make sense of it? Is it meant to be made sense of? Some refuse to view it as something other than suicide, some refuse that simplification. The "who" was the thing that kept me holding that image of Bushnell from the blurry photographs. This meditation on his act was the only way that I felt I could honor what he did while keeping it separate from any acts I might undertake to protest the same problem. I could only hold this image, rather than drop the too-hot coal of it, by wondering who else had unleashed this most spectacular protest, by learning what embattled lineage he joined by that act.

In a chart of self-immolations that have made the news over half a century or more, my eyes gravitate toward the American names—like a truly warped Olympian accounting. There was

Alice Herz, who came to the US to escape the Nazis but later found LBJ sounding the war drum in Vietnam. She was the first person known to self-immolate on US soil in protest of the Vietnam War, and did so once she felt all other modes of protest had been exhausted.[30] Many United Statesians followed suit thereafter. There was Jewish Columbia student Bruce Mayrock, whose self-lit body blazed before the UN building in New York in 1969 in protest of the Biafra genocide, and whose body was finally extinguished by UN guards when he dropped to his knees next to a statue with the biblical words "Let us beat our swords into plowshares" inscribed on it. There was Atlanta's Willie B. Phillips, who, in 1972, at twenty-seven, said, "I'm tired of this old world" as he doused himself in gasoline, then set his body alight in protest of US racism. There was Antoine Thurel, who set fire to his body in front of the Massachusetts State House in protest of the US's backing of Haitian dictator "Baby Doc"—a newspaper clipping from the day shows that Capitol Police "thought the 7 a.m. self-immolation was a trash fire."[31] In recent years, climate activist Wynn Bruce self-immolated on Earth Day in front of the Supreme Court building.[32] He was said to quote Thích Nhất Hạnh in a Facebook post before the act: "The most important thing, in response to climate change, is to be willing to hear the sound of the earth's tears through our own bodies."[33] An unidentified person burned in front of the Israeli Consulate in Georgia in protest of the genocide in Gaza.[34] An unidentified male burned in Kinshasa, Democratic Republic of the Congo, holding a sign that said "Stop the genocide in Congo."[35] Combustion does not only erase ("burn after reading"), it can also preserve a message (as in pyrography). Fire makes the news. That's why the desperate, the fed-up, the enraged, why those rising up wield it.

I hesitate before the word "martyr" today at the protest, my Western tongue tentative around its ancient syllables. What if those who burned so quickly had used their breath to oxygenate these fires instead? Is their cool breath more help in death? George Sweeney, writing on political self-sacrifice and citing Hocart, says, "It has been argued that religio-political based self-immolation occurs in societies where the dead are given higher status than the living."[36]

On the podcast *Mushroom Revival,* mycologist Monika Fischer spoke of the *Pyronema* fungi that are born in the crucible of fire, which break down burned organic matter and then die themselves, as a kind of gift: "I love the hypothesis that *Pyronema* is basically gifting its body—its biomass to the organisms that come after it, [which] is really easy food for any organism that's going to come after it." Such a process, she says, could slow erosion.

In a since-deleted Facebook post, Bushnell wrote, "Many of us like to ask ourselves, 'What would I do if I was alive during slavery? Or the Jim Crow South? Or apartheid? What would I do if my country was committing genocide?' The answer is, you're doing it. Right now."[37]

Our breath in the streets, slogans spilling from our lungs. How our breath bellows at its task. How we link arms in a human network, burning visibly, how we are the metaphors for them.

IN SEARCH OF

Tricholoma magnivelare

(THE MUSHROOM AT THE END OF THE WORLD)

On portals and where they lead, work and money, property and theft, the end of the world as we know it (do you feel fine?)

Never forget that work is the story we tell ourselves about money.

—Eula Biss

"Mushroom nerds are like the only demographic of human where you can drive to a random state to meet with a total stranger from the internet alone in the woods and just geek out and have a great time and eat delicious food," says Matt G. when I ask him about his time as a self-styled itinerant mushroom hunter. "I ended up in Canada and Alaska. Central America. A lot of the East Coast. And would hunt their best spots with [folks] and learn a bunch of new species and share things others had shared with me."

Minus the driving, this description of plain and earthly heaven, of far-reaching mycelial networking, belongs in a holy book. It was supplied to me by mushroom nerd Matt, a beloved mycological meme-maker and ecology buff on social media who, and I can't explain it, is the physical embodiment of the memes he both makes and aggregates about mushrooms, politics, and our responsibilities to the planet and each other. There is a spirit of the jester about him even though his delivery is bone dry—as a former class clown myself, I recognized him as kin from across the country. This allowed me to trust him to be my very first "total stranger from the mushroom internet that I actually meet" when I heard he was going to be in the Boston area several years ago. And while it was incredibly dry when he visited that September, we still found enough in my accustomed range that was "edible" to make a truly edible feast together, which we shared with my partner's co-op. "Edible" is Matt's tongue-in-cheek catchphrase, an understatement for describing good food, and an oblique reference to the culture of online mushroom identification forums.

Matt found life beautiful when he lived to follow mushrooms. "Chasing the rain was really fun, as was using social media in an actually social way." A combination of savings and spoils from the economies that spring up when people are in true community funded his primary expenses of gas and meal supplies. "I had a lot of freedom of movement and could stay and go on a whim. What little money I made here and there seasonally/nomadically stretched way farther in terms of enjoyment of life."

When I ask whether he'd ever picked mushrooms commercially, Matt says not really. "I'd share them more than anything. It's most fun to pick a bunch and make a big collaborative dinner with friends. Though in Alaska I was really broke and needed commercial fishing gear and a fishing license, so I was selling ounce bags for twenty dollars a pop because I had so much abundance. But I was also sharing them and trading them."

It wasn't just for the bare necessities, either. "I did also use dried mushrooms to barter with artists online. I'd see something they had made that I loved, and I'd message them to see if they wanted to trade. Some people wanted money, others were delighted. Pretty much all the art I own came from trades like that. So, I guess [my ethos was to] share normally, barter with strangers, and sell when completely broke."

Living is more difficult now in his settled work, which he only does in order to pay rent. Not a week goes by when Matt doesn't share with his followers how acutely he feels the difference in these two ways of living. "Now that I work I rarely go out. I have a core set of friends that I established when I was homeless who I do hang out with and basically do all the fun culinary shit we like, but I haven't made new friends in forever. I'm usually tired or stressed after work and just want to decompress and isolate because I don't have the energy left to expend on other things."

It's easy to forget that other ways of living exist when the most stultifying ones have the clearest, most official roadmaps. But portals into living your interests abound. And walking through them can be as easy as coming down from human scale to put one's knees and cheek directly into contact with the Earth.

Anthropologist Anna Tsing writes of these portals. She writes about how they are often active at the edges of precarity, where we cop to our inter-species collaborations, where we're vulnerable to and contaminate one another. She writes of how "indeterminacy, the unplanned nature of time, is frightening, but thinking through precarity makes it evident that indeterminacy also makes life possible." She writes of how "as contamination changes world-making projects, mutual worlds—and new directions—may emerge."[1] A new world is made when we collaborate across difference.

Staghorn sumac leaves erupt like prayer flags through a fence along the highway, sunrise-brilliant and animate on the wind. Billy, a photographer of the more-than-human and their allies, is driving us south in his trusty Camry, and as he waits to fully merge into 93's southbound traffic, I take advantage of this slowness to watch for deer and mushrooms in the banked woodland nearby. For a perpetual passenger like me, this game has made the world that races along the periphery a companion, so that I know the highway landscapes in eastern Massachusetts more intimately than I know many of its neighborhood streets. Sumac still abounds in the margin once we've picked up speed and I wonder what mantras the plant would send into the ether, the way prayer flags are meant to. As if whatever I find will answer this question, I google what sorts of fungi partner with plants in the genus *Rhus*. Unsurprisingly, knowing the names of the microorganisms that are likely paired up with the roots of these roadside shrubs doesn't tell me what sacred messages they might wish to convey heavenward, but it does let me speculate my way into the worlds at their feet.

On the highway, most of the traffic heads in the other direction, northward toward Boston, while our destination is Cape Cod. The traffic is workbound, and in certain ways Billy, who is never not thinking of ways to make his next photo documenting local ecologies, is too. Because I will write about the *Tricholoma magnivelare*, the famed matsutake, the mushroom at the end of the world, that we're on our way to pick, this could be conceived of as a work trip; if I were the one driving, I would certainly be saving gas receipts for tax purposes. But the act of tracking fungi themselves is only work when an organization or individual pays me to lead educational walks (which I also do for free)—disseminating mushroom knowledge feels like renewing the resource rather than spending it. I have never assigned a dollar value to a fruit body. This is more about my temperament and the abundance of the land I live with, how nice it feels to give the bounty away, than about any principled abstention. That said, I know it would feel different to bring scales and expectations into the woods with me. As people who make art, Billy and I head out into the oak and pitch pine forest for activity that looks a lot more like play than work.

The mushroom at the end of the world—I love this epithet Tsing has gifted us, its theatricality—the way it expands in meaning as you hold it in your mind. I love it as a handle for the matsutake, or pine mushroom, an object of beauty, longing, and desire for centuries in Japan and lately an object of fascination for me. As edibles, matsutake are polarizing in the US. My own partner dislikes them with the zeal of one who has tasted the truth, and he takes every opportunity to make that known when we're among mushroom people. His response makes me wonder if there's something genetic at play, like with cilantro. The fungus's scent was famously described as "a provocative compromise between red hots and dirty socks" by David Arora long ago, though I wonder how he feels about that pithy descriptor now. I prefer the more palatable "aged hard cheese and the freshly peeled bark of a cinnamon tree." But we're all essentially saying the same thing, whether we're singing a love song or not—these fruits are sharp and spicy. Your first whiff of them will be a confrontation.

Billy starts to fill me in about the latest in his life and we pick up steam, the traffic clearing. If you look at a map of Cape Cod and imagine the spit of land is a flexing arm from shoulder to fist, we are headed to the wrist. Provincetown, a land at the end of the world, is the peninsula's fist. It's also a summertime camping destination for my friend group and an LGBTQ haven; gay men dying during the AIDS epidemic having found community and quality healthcare in Massachusetts and a soft spot to land when their hometowns grew hostile.[2] Provincetown is also where the

Mayflower landed in 1620, starving colonists stealing the winter stores of maize stocked by members of the Nauset tribe who were away at their winter hunting grounds, then leaving a note in English that they were good for it.[3] A great example of early white American foraging.

On the peninsula's wrist, in the previous October, Billy and I found five beautiful, edible matsutake fruits, along with a couple that were better left for bugs to enjoy. Five mushrooms may not seem like much, but it's a treasure for someone who had only ever encountered matsutake in other peoples' spots or growing singly, weathered and bug-eaten, on the verge of rotting. We are hoping that with the entire day to play with and two known spots, our chances are greater. No matter the circumstance, whether I would be able to eat what I found or not, the rush of finding matsutake is unrivaled in my experience; I am speaking to you of one of the most intense joys of my life thus far. Last year Billy seemed to catch the bug secondhand, excited to track what he called ground potatoes. I suspect much of his enthusiasm was about the possibility of capturing the payoff on film, since he didn't take any of the mushrooms we found home, though I tried to insist.

Billy first contacted me near the start of the pandemic when he heard a public radio story about my taking Black and brown femmes into the woods on mushroom forays. He said he wanted to take pictures and perhaps make a story about me. I was flattered; his photos are mesmerizing. Among my favorites when I searched his online presence to gauge how likely he was to murder me were the black-and-white portraits of women in Massachusetts who practice falconry. Billy has since ferried us to several locations to document the hunt, including to disused apple orchards and abandoned pig farms to find morels. He's even brought me treacherously close to becoming a birder, like the time he invited me to watch and count hawks in a cemetery tower. Over the course of this association and a couple of small collaborations, Billy and I have become friends. Which is lucky because this day, we don't find any matsutake, and he doesn't seem to take any photos he's happy about—still, time in the woods with a friend is never wasted. Plus there are plenty of red-capped *Leccinum* going home with us.

He isn't the only William I've day-tripped with on the trail of what Japanese poets call the "autumn aroma." About a year before I met Billy, Bill N., a veteran of the Northeastern mushroom scene and a brilliant teacher, told me he would make my matsutake dreams come true. In a series of Facebook messages that make me smile to reread, he promised to bring me along on one of his upcoming matsi hunts in the mountains of New Hampshire. I had encountered Bill in person once before, very briefly, at a meeting of the Boston Mycological Club. He'd lived in Jamaica years before and was eager to demonstrate his familiarity with some of the finer points of my culture. As much as I've since joked about getting into a car with a stranger who always carries a knife, whose idea of a good time is waking up early and prowling a woodland he knows down to the humblest twig, even though we were merely acquaintances who tried to out-dad-joke one another on the internet, it didn't feel wholly right to call him a stranger.

To be flanked by the White Mountains at the height of the leaves' changing color is to be bathed in the complexion of Earth at her most hallucinatory. The bald granite, the evergreens, the uninterrupted technicolor of it—you could fall away from detection forever and practice freedom here, as the license plates are continually admonishing you to do, *Live Free or Die*. On the day Bill and I drove along the 302, winding our way through the valley corridors, it felt like passing over threshold after threshold. Jeff Beck played on Bill's car stereo, a special mix created for road trips. He's chatting about the online mushroom community—the differences between we younglings coming up through it and the way networking used to be done at regional forays. We talk a little about spotting mushrooms from the car, how sometimes the most impressive hauls come from spending the day driving suburban and rural roads, especially for quarry that forms larger fruit bodies. Mushroom topics are inexhaustible, and there's even room for a little gossip about other veterans in the community, younger and older. The messy public breakups; who has a terrible memory for Latin binomials. While we chat, I'm looking through photographs of the pine mushroom that I've saved to my phone. It's important to get the search image ready in my mind's eye.

Here are some other, slightly more boring things to which you might compare finding a flush of matsutake mushrooms: Winning an Easter egg hunt set up by the adult who refused to patronize you children by hiding the eggs in plain sight. Hitting the *x* ring five times in a row during a round of archery, then suppressing your smile under a mask of neutrality when your companion reacts. (Bonus if you've ever pictured this companion naked.) Realizing that a comedian who has made you cry-laugh till your stomach hurt and you did the high-pitched inarticulate laugh-talk thing in an attempt to repeat their joke has four other specials ready to stream. (Bonus if the comedian hasn't been canceled for being a sex pest.) Finding the perfect gauzy dress at the thrift store for an upcoming occasion, and not only does it fit like it was tailored for you, but it's well made and priced like the hipster who works there never saw it. Stumbling upon the perfect sexy movie to watch with a fully charged vibrator and the next two hours home alone. Happening upon shed moose antlers, which is not something that has happened to me yet, but on the day it does, trust my delight will be complete. Finding a $100 bill on the street.

Bill and I have moved on to a second location and are driving toward a trailhead when he tells me to put my eyes on. He says this as if they haven't been on since I got into his car hours ago (we've made a few stops, including a restaurant where he sold some matsutake that morning, and a late chanterelle patch where he talked with an insane-seeming man about local politics). To have spotted a worthy terrestrial mushroom from his moving vehicle would have made it clear I mean business, so of course they've *been* on, but I make a show of putting my nose right up to the window, nonetheless. When I see them, their wonderful wee domes, like someone has poured au jus over a perfectly poached goose egg, I'm sure I vocalize. He stops the car. He is also a mushroom hunter, a mushroom lover, a mushroom photographer, and mushroom educator, so he understands that we will be here for a while as I examine these stunning specimens from every angle. They would get a B grade on the market—their veils haven't dropped but they are no longer buttons. Bill goes into teaching mode.

"Their specific epithet is *magnivelare*: 'magni' meaning great and 'velare,' 'to cover' or 'veil.'" He mimes "great" with his considerable wingspan. I appreciate the descriptiveness of the Latin, and that he's choosing this moment to share the knowledge since he knows the adrenaline will lock the words in my memory. I *can't* remember, however, whether I give an intelligible reply or just continue squealing and sniffing. And this is not my first encounter with that famous scent, either. Bill had arrived to pick me up with baskets full of matsis for the restaurants he supplies, which perfumed our entire ride. But it is a subtly different pleasure to inhale their essence fresh from the ground.

We go on to several other locations, wending our way through lakes and riverbanks, at one point crouching low to avoid detection by dogs as we pick at the edge of someone's property. To look for these mushrooms in their grade A state means being alert to slight bulges tenting the hemlock duff, which mushroom people know as shrumps. As the day progresses, I get better at spotting the right shrumps and disinterestedly using my foot to clear the needles off the "wrong" ones, as some of the tents are made by the spicy *Lactarius piperatus*. The presence of these others can fool you at first, but they can also indicate that you're in the right environment for matsutake. Peppery milkcaps, as *L. piperatus* fruits are commonly known, also present a great opportunity to annoy a friend: Just dare them to taste and spit a little of the flesh of a fresh-picked specimen, watching as they realize just why it earned that common name—which of course you reveal to them only after it's too late. Bill and I pick past twilight, ending the day next to a darkened stream, its whispering susurration a natural muscle relaxer.

As we head back toward Massachusetts in the dark, I admit, "I really thought you were going to blindfold me when I first got into the car—you've taken me to all these spots that supply you with your livelihood."

He replies, "It doesn't matter that you know where they grow. I've known the organisms themselves for years. You couldn't really beat me here." I was only teasing, of course; he knows I know the etiquette of never going back to a spot you've been invited to without the inviter. It's something like the rules that govern vampires and thresholds—like a magic disappearing portal. I've never been back to those locations, though they're geocached on my phone.

A good friend of mine makes small-batch vermouth in the Hudson Valley using ingredients she forages. Her brand is known in the vintner space—for a while hers was one of the only vermouths made Stateside. I met her at another friend's wedding at a riverside park in Brooklyn as a klezmer band played and the Williamsburg Bridge loomed grand as a mountain in backdrop. B and I clicked instantly after having found a corner to smoke weed in—then danced together like no one else was there for the rest of the ceremony. B was cool and direct and cut through the bullshit. Her presence is elemental. For that reason alone, I'm glad the evening of the wedding I had a "chill with light paranoia" high and not a "panic attack" high.

As a forager who spins the ingredients she finds into a commercial product, her relationship to what she reaps but did not sow is different from mine. "#foragersarethieves" is a hashtag she appends to the wild food photos she posts on Instagram. "Foragers are thieves," she declared during the Q and A for a panel at a spring festival near her home, where I'd been asked to speak with other foragers about my practice. In the latter context, I understand the impulse to take the proceedings down a couple pegs—it's a little perverse, if you really think about it, how much finding your own food has been romanticized by those of us who are alienated from the food systems that feed us. It makes some sense to wonder out loud what precisely it is that we're doing when we treat the act like some fringe, quaint custom when many people around the world have never stopped eating this way.

But I keep thinking about her use of this slogan to sum up an activity that has never felt like stealing. Especially since the phrase isn't clearly about reckoning with our settler status. Having always instinctively agreed with Proudhon's notion that property is theft, with what indigenous people around the world have always known, that you can't own the land, not really, this idea had me wondering: Who are we as foragers stealing from? If you believe, as I do, in a person's basic right to feed themselves by virtue of being born, then it doesn't really follow that picking mushrooms in a public wood could possibly count as theft. But this sense of what is legally "ours alone" is what organizes our lives in the West, and while I'm sensible to the distinction between billionaires building apocalypse bunkers on newly bought islands and small business owners providing goods and services in exchange for their own future and security, I could never embrace any logic that calls picking nettles in the spring amoral. Or maybe I just prefer that my career as a thief gets the past-tense treatment.

I ask Matt whether foragers are thieves, and his answer is predictable in substance but surprising in his thoroughness. Where I'd been expecting simply "no," he says:

"They can be depending on the context and ideological belief, or not. I think walking through the woods and eating food off of the ground is fine. It's however only legal in like three places in California [where he lives] to do this. The people who made it illegal were plant societies and plant society lobbying power who saw the commercial mushroom hunting trade becoming a powerful entity in Oregon and Washington and made it illegal in California. Most plant-people ideology is based in a post-genocide museum-under-glass ideology that deliberately and intentionally ignores generations of human coevolution and anthropogenic food-space shaping and substitutes a stay-on-the-trail and leave-no-trace mantra. Which is stupid and racist. It also ignores basic fungal reproductive biology which is also very stupid. Plants aren't fungi, picking a mushroom doesn't harm the organism underground. And also their goal of stopping commercial hunting in California through these laws only made it easier for commercial people to get away with it while hurting subsistence foragers and also taxonomy nerds. Foragers are thieves if you believe that geopolitical bureaucratic entities can own land and dictate human behavior on it. I'd say politicians and landlords are thieves before I'd say foragers are." A whole sermon, Matt, go off.

Does the calculus become different when mushroomers are picking commercially? Anna Tsing writes that "matsutake picking is not 'labor,' but it is haunted by labor. So, too, property: Matsutake pickers act as if the forest was an extensive commons. The land is not officially a commons. It is mainly national forest, with some adjacent private land, all fully protected by the state. But the pickers do their best to ignore questions of property. But what is 'public property' if not an oxymoron?"[4]

(If you don't know about the hot mess that is "the tragedy of the commons," a concept popularized by an all-too-popular essay arguing overpopulation would necessarily lead to the exhaustion of Earth's resources, based on the mythical idea of land that was open to all yet somehow completely unregulated, an essay that spawned hundreds of pages of better-informed rebuttal . . . then check it out on Wikipedia. It's a good time.)

What does it mean to prohibit foraging, to exclude people from the land in one of the most basic ways humans have always related to it? Jumana Manna's 2022 film *Foragers* mixes fiction and documentary footage to highlight the answers to these questions.[5] It follows the stories of Palestinians who have fallen afoul of their Israeli occupiers' prohibitions against the picking of a vegetable called 'akkoub and the zesty herb za'atar, incurring heavy fines when caught in the act of picking, attending court dates in which they make impassioned declarations and defend their customs, citing the necessity of feeding their children. The character of Wardeh explains that since these nature protection laws have penalized people for interfacing with the land as is their custom, the plants aren't being sufficiently taken care of. "Since you banned us," she says, "za'atar is harder to find. Za'atar needs to be trimmed, like all plants. The more it's clipped, the stronger it grows back." Another warns, "I'm waiting for this to be over so I can go back to foraging. . . . I'm not paying a penny. . . . I won't justify your law. . . . This law is shit, banning us from foraging food. What protected species? The land isn't yours. Neither is the plant!" The unseen interrogator lists all of his offenses, naming the years when he got in trouble. A little over a year after the film's release, and nearly four months into the genocide in Palestine, I watched in my living room as Samir, who plays himself, made a rejoinder that forced the word "Yes!" from my throat: "I'll also be caught in 2050 with my children and grandchildren. I'll continue the path of my grandparents. That is my truth."

Nothing says "outdoorsy elder millennial" like the Subaru Outback my partner and I share, with its kayak and bike racks, its rock-climbing crash pad stashed in the trunk alongside two hammocks, a small, two-person tent, those battery-powered mosquito repellent devices, two different camping stoves, LifeStraws and water purification tablets, a bear canister, and a National Parks pass in the glove box. When we first purchased that last, it felt like some kind of ridiculous graduation for the two of us, who had never owned a vehicle before the pandemic, to suddenly have readier access to so much of the US's natural beauty. Looking into the history of Shenandoah National Park complicated that feeling.

Shenandoah is an accustomed stop for us on our annual road trips to and from Florida. I remember being excited, during my first visit, to see official acknowledgment that people will be foraging there whether it's sanctioned or not—as of this writing, the park allows one gallon of morels to be collected per person per visit. Looking up the land-use history before embarking on my first hunt on the grounds led me to reading about the Lewis Mountain camping area, a designated "Negro area" within the park from 1939 to 1947. There were integrated spots throughout the rest of the park, but where they were was not common knowledge. It did not take much further digging to learn about the world's first National Park at Yellowstone, how the US military systematically ethnically cleansed the area of the Indigenous people who had been making use of it for millennia. Nor was it hard, despite the National Park Service website's stilted writing on the matter, to find where the westward expansion of America was aided and abetted by new ideas of conservation and environmental protection, ideals that conveniently sprouted up alongside settlers' continual impingement on Native lands. The US military's protection of the National Parks was at the forefront of this war, as my adopted country championed a "museum-under-glass" view of nature as a separate entity, as "pristine," "unspoiled," "wilderness," pure and Edenic, which must remain untrammeled by people, so it can be looked at.

In light of all that, what does it mean for an $80 rectangle of plastic to grant such sweeping access to lands from which Native peoples have been systematically excluded, people who are actively fighting to get that land back "from sea to shining sea"?

The most capital* T *thieving I've done took place when I was twelve and was caught stealing press-on nails, brightly colored hair accessories, and lip balm at PharMor pharmacy. Once I was unceremoniously deposited in a back room by a strip mall cop, another man, who was probably a manager and definitely long-suffering, kept reiterating that the company had lost so much thanks to *people like me*. I did not know then that PharMor's founders had been embroiled in one of the biggest corporate fraud scandals in American business history, stealing to the tune of $500 million from shareholders.

After some initial questions the store employee asked to get all the info he needed to bring my mother on the scene at the same time as the police officer who would file the report, I don't remember feeling a thing. I'd been emptied of any sentiment once it became clear that we were going to go through with the great theater of Enforcing the Law because I'd helped myself to fifteen dollars' worth of plastic.

Being caught shoplifting. What a shameful way for a Black girl to behave! Not because to shoplift is to breach a social contract I wasn't old enough to sign, anyway. But because I had summoned the state into our lives and in the process frightened my mother. *Being caught.* This was *so* not how Michelle Trachtenberg's Harriet the Spy, whom I felt I could have snuck about with, would conduct herself. We'd have neutralized the surveillance cameras together, somehow, first; made fools of these adults under their own fluorescent light.

But for my mother—how terrifying that her middle child would convene this chaos of cool flashing lights and reminders of what could be held against us for a mini-backpack full of trifles—that this is what I chose to do with my recently naturalized life, my burgeoning freedom. My mother couldn't have known that I didn't actually want the things I stole, not in the way I was supposed to want them. My mother couldn't have known that the small, secret collection of feminine items under a pile in my closet, which couldn't be worn or used anywhere she could see me wearing or using them, were only, perhaps, a symbol. She couldn't have known that what linked my brain chemistry to one-quarter of the prison population's brain chemistry is the dysregulation of dopamine.[6] She couldn't have known whether I would continue to move through the world boldly displaying the target on my back like this.

I like to believe I'm clearer now about when chasing a feeling is amoral, but who knows. I won't, post facto, try to valorize my dirtbag, tweenage, thieving ways. Besides, I can't actually remember to what extent I was saving that dragon's cache of petroleum products for a hoped-for femme debut. Maybe I'd dreamt of a different version of myself than the one who wore overalls and flannel in the Florida sun. Maybe I thought someday I would steal one of my mother's sexier shirts, toilet paper rounding out my training bra (training what for what?), high ponytail spilling over with hot-combed, Ultra Sheen hair, press-on nails tapping the desk in Mr. Dowling's sixth-grade geography class, restless and ready for the bell. This version of me is, perhaps, ready to stalk the schoolyard for boys rather than head home to read *Goosebumps* books until her eyelids give out. Hunting for boys would have been much simpler than hunting for portals into another world where I felt more real.

I'd seen glimpses of these portals, so many when I was a kid, and they would continue to haunt my senses along with the objects, open spaces, and peripheries of each of my past lives. On vacation. In the stock-still gold tractor beam of noontime sun coming into my room, just before an illicit step outside on a sick day. When an acquaintance lived on a houseboat, never docking. When mutual aid materialized in the wake of a storm. In a march from the Democratic National Convention to the Republican National Convention in protest of both. After a death. After many deaths as in a pandemic, when folks got addicted to quitting. When former classmates ran away and couldn't be found again on the internet. In tents on Wall Street. During a riot, looting, an uprising. After a bombing when marathoners kept running, now to blood banks to donate blood to the injured. In a power outage. After severe weather as we *used* to see it—innocent of Earth's course-correction. When we're lost and looking toward nature to get back to where we should be. It's a start. As long as we remember that this does not mean to cast our gazes "outward" toward something else.

When I first began to post photos of mushrooms I thought were cool on social media, weekly, sometimes daily, an acquaintance noticed. "Do you work?!" she asked on a photo of a blue-staining bolete or a quilted russula I'd added to the album I'd named *I Told This Arachnid It Had a Beautiful Body and Now We're Engaged*. She works in real estate. I felt defensive before I thought about it. I was "underemployed" then and always had been. I walked and trained dogs, sometimes publishing short stories while working on novels and living rent-free with a wealthy partner, so yes, my days were largely my own. My defenses were up because I was used to feeling guilty about having all that time to dispose of as I wished and for using it to do useless things like photographing the smallest parts of the day. Nobody was paying me to indulge my fungal curiosities, and nobody ever would. I was exercising freedom I didn't really have, given my college debt. I wasn't at a desk. I wasn't saving for retirement. "Do you work?" sounded like "Are you responsible?" "Do you think you're better than me?" "Have you even tried to get a real job?" "How will you ever know what your potential is if you don't even try?" "Don't think you can hack it in grad school?" "Isn't it time to stop pretending the writing will take you anywhere?" "Why, precisely, haven't you taken up the mantle of adulthood?"

As l'esprit de l'escalier would have it, I wish I had said that no, I don't work. Like any freelance artist in the twenty-first century, I *have a practice*.

Now, the next four months are stretched out before me, an expanse of hours moving from winter to spring, and the plan is to write and read and sometimes talk to people and sleep and eat and take walks in the woods and that's it. I am taking this time for myself to see whether this package of questions and bewilderments and assertions and self-definitions and unearthed histories can come together into art. This defiance is age-old. I am taking freedom in a way that someone somewhere will say is stealing—stealing from the people who gave me those loans, stealing from my future according to good sense, stealing from some nonprofit who could use my underpaid labor for their aims, stealing from the civitas. This will be the most selfish four months of my entire life thus far, or the most self-styled, anyway. I have no guarantee that money will be on the other end of it in any form—these movements across the page are not a commute. I've recently been having frank discussions with other writers about publishing, about what a writer like me can expect (it starts with a *p* and rhymes with "menury") and what she can't (to be paid). If there's any actual hope, it's that perhaps someone will bet on me again because they see potential. This is freedom. It is reckless and I didn't do anything to earn it.

It's February, and my freelance income stream is more of a sluggish trickle. In this state it's hard to imagine "summer's easy riches," which had found me in possession of a bag brimming with *Tricholoma mesoamericanum*, the Mexican matsutake, at a technicolor morning market in Oaxaca.[7] The fog had burned off just enough so that only the Trader Joe's–sized expanse of vendor rows existed—all the rest of the world stayed vague. The bag of matsis cost 500 pesos—around $30. "A rudely yawning one" is how I would describe the chasm between 930,000 yen ($6,235), a sum that fetched less than half a pound of prime matsutake at auction in Japan last October, and the thirty bucks I've just paid for these Oaxacan fruits in the same genus.[8] Back home in Massachusetts, my friend Tyler, who owns the Mushroom Shop in Somerville, sells grade-A or no. 1 mushrooms at $50–100 a pound, depending. So much depends.

Our vendor is a Mixtec woman in a puffer jacket and camo cargo pants on a small wooden stool. Before her is a rainbow of perfect mushrooms—yellow corals, reddish orange *Amanitas*, fuchsia *Laccaria*, peachy *Hygrophoropsis*, and some black, glossy polypores, the likes of which I'd never seen before, separated neatly on a tarp. She smiles as I bury my face in the matsutake bag, filling my lungs with their scent and spores. Any travel weariness dissolves. I detect something like the same cinnamon tree bark of the ones back home, that deep umami, and yet there's something else there—maybe rain? The only thing that could improve this moment is if I had found these beauties, known locally as *ji'i yisi* or *hongo de aguacate*—if I had spotted them forming shrump rings in the fogged pine forest where they grew, having read the woods well enough to get to where they'd assembled, right as the first rosy fingers of dawn touched down around them in the duff. But on a first visit to a region, that's not the likeliest of outcomes, and this feels like the right way to encounter them today, so that I'm a participant in the local economy rather than an extractor from it. (Can foragers be thieves?)

Later that morning at a café where the chef will cook some of the mushrooms we've acquired at the market, our host will talk about the influx of outsiders looking for matsutake spots in the cloud forest and the local hunters' practice of keeping mum. The group I'm with consists of Ñuu Savi land stewards and organizers who lead mushroom tours and conservation efforts in Oaxaca; a mycologist and professor at Universidad Nacional Autónoma de México; a mycophilic political science major from the area and her French partner who helps translate for us; and a politician, his mushroom-enthusiast wife, and their myco-prodigy eight-year-old (the soul of our group). We're gathered at the table, chatting about mushrooms until that topic is exhausted . . . and then someone finds a way to bring the fungal back round again. Nerds gonna nerd out, in any language.

"Ñuu Savi" means "people of the rain." Since that trip, I've begun mentally assigning regional characteristics to anyone who protects the land—the Cape Codders who help fight erosion are people of the dune, for instance, and the Jamaicans who fight bauxite mines in Maroon territory are people of the star valley. *Foragers are thieves.* Ñuu Savi is the loveliest two-word poem I know at the moment.

When our mushrooms have been roasted, we pass them around—the brilliant blue *Lactarius* having faded to a tawny green on the plate, the hongos de aguacate now radiating a scent more like old cinnamon sticks from the back of the cupboard. One mushroom we bought that morning was particularly striking in its size and color (large, yellow) and was referred to as "cow's tongue" at the market. It's my favorite taste among what we share in this lightly salted, nearly fatless preparation. Later I'll see the same mushroom referred to as *panza de toro* (bull's belly) or *jiyii yaa'stiki*. When Carlos, the mycologist, talks about how locals won't divulge matsutake locations to outsiders, I'm reminded of Marvel, of all things, and of the fictional, ages-long cone of silence around vibranium in *Black Panther*—the way wealth can be a practice of communal care, of protection, of not naming a price even to the highest bidder.

The conversation turns to the ironies of ecotourism and the differences between being in community in the places one visits and touring in the traditional sense. Carlos, who is visiting from further north, is technically a tourist but undoubtedly in community. My companions speak too quickly for me to participate, both for language reasons and also because I'm distracted by the lines from Jamaica Kincaid's *A Small Place* that have been swirling in my mind since I got to Mexico (silly because Mexico is a large place, but my mind has its reasons), and I'm not clear whether I'm on the right side of the trad-touring and conscious-touring divide. The lines are, "This might frighten you (you are on your holiday; you are a tourist); this might excite you (you are on your holiday; you are a tourist)."[9]

I am one of those people who can understand Spanish depending on the speed of the speaker's tongue and the novelty of their accent for me. I can speak a small child's Spanish fluidly only if I let go of perfectionism long enough, which these days only happens when I'm intoxicated. There are no cervezas at this breakfast, but I have brought exhaustion and altitude sickness. They don't work the same. When the conversation switches to cultivation and medicinal mushrooms, Carlos and our translator happily loop me in again without my "¿perdón?," perhaps assuming I'm less interested in the former topic than this one. The reverse is true. Nevertheless, I talk about my use of lion's mane tincture and follow Carlos on Instagram, where he posts photos of his grows and hunts alike.

The next day, we pile into our host's pickup for a long drive from the cabañas where we'd been staying, on a campus that's pretty empty save for an armed guard, a receptionist, and a semi-friendly German shepherd, off a street named for Porfirio Díaz. We're headed to a fiesta de hongos in its lucky-number seventh year, and will drive three hours southeast through the Sierra Madre del Sur mountain chain.

Upon arriving in the woods around ten in the morning, I hear trumpets positively embossing the air, announcing that this will, in fact, be a party—that it's not "fiesta" in the same puzzling way US folks can use "party" to refer to a gathering where all you do is have a nice dinner. Listen, in my home woods, if I hear loud music, I want to throw hands. That's because it generally interrupts my peaceful walk, via the utility belt Bluetooth speakers of Serious Runners or Mountain Bikers who are terrified of the sound of their own thoughts. Its arrival, in that context, is unpleasant for everyone but the individual who brought it.

But this mariachi brass is holy, its effect on the diffuse and shadowed warmth of the clearing before us sacral. The browns and reds of the duff are deeper with their sounding, the pines loom and sway in their rhythm. The smoke from the mobile cooking setups of food vendors is holy smoke as it rises with their trill; the way kids are dancing and darting about, teasing elders, how deep into the forest we had to drive to find this sound, this is all evidence of the divine. Church on a Sunday. As we approach, I catch the word "hongos" in the microphone-free clarity of the singer's voice and the delight I feel at hearing "fungus" valorized in a corrido has me dance-walking toward the picnic tables, where we'll set ourselves up between a couple of families. The scent of roasting mushrooms, here again. Alongside that familiar aroma is a sweetish brew I can't quite put my finger on.

"¡Que vivan los hongos, que viva la Ruta de Hongos Mixteca!"

When the band is finished with this song, they talk to our hosts and offer us alcohol from a plastic water bottle. The hour, my worsening altitude sickness, and some sense that whenever I'm being offered alcohol as a woman outside of my usual social milieu, it's a test, impel me to start to put my hand up in a gesture of refusal. My partner does the same. But our hosts accept the bottle and pour for us, anyway, knowing that we drink because we did so gleefully during the mezcal tasting the night before. This homebrew is fruity, weak, and pleasant. The rest of breakfast is a sampler of pozole, tlayudas, tamales de hongo, and tejate. The latter is the only thing I had not tasted before and the only one for which I have no point of comparison. Known as the drink of the gods, tejate is a mix of toasted corn, toasted mamey seeds, cacao tree flowers, and fermented cacao beans, made into a paste that's massaged for hours, with water added little by little. The resulting sweet, slightly waxy beverage is solid and liquid both, a little earthy, pre-Hispanic, and wholly unique.

After breakfast and a rousing invocation delivered by the event organizers, including our hosts, we're divided into teams and loosed upon the forest with our canastas and instructions that are the same in any language: when you find a mushroom you're not familiar with, leave it where it is so that an identifier can examine it in situ; don't collect every edible fruit body you happen upon; sometimes a mushroom requires the excavation of material below ground for proper ID. We're also told to return fungal matter we have no further use for to the soil by burying it. The burial directive was not one I'd been given before, but I planned to incorporate the practice into my own walks moving forward. *Foragers are thieves.*

Once we've spent a couple hours collecting and cooing over the conifer woods' gifts, we bring the mushrooms back to the table for identification. Not one to let the opportunity for a speech pass, our host delineates the long mycophilic tradition that the event is a part of, going back centuries, as depicted in the Mixtec codex Yuta Tnoho. This text illustrates the creation of the universe, among other things, and details mushroom usage in ritual contexts going back to the fourteenth century CE.[10]

Speech over, Carlos gets behind the table and reintroduces himself, asking some elder women to join him there as he describes the specimens found during the foray. We've been instructed to put edibles on the right side of the table and nonedibles on the left. Carlos is at pains to honor the place-specificity of the event—if a mushroom is eaten elsewhere but not in this region, he does not emphasize its culinary uses. The care that is taken to situate the proceedings both in the contexts of the past and future, with nods to tradition and sustainability (I believe I heard something earlier about protecting this forest from development) makes a foray that started out feeling holy end feeling whole. Later, when I praise the inclusion of grandmothers at the ID table, Carlos says that these women are mycology's protagonists, so of course they should be included at every foray table. It's this collaboration between the traditional, the academic, the community-building, the commercial, the free, the play, the work, and the citizen-science that leaves me so inspired.

At the beginning of our ride back to Tlaxiaco, through the forest and before we hit the highway, Walter opts to stand in the truck bed with our poli sci student, Diana, both of them holding tight to the vehicle's frame over the bumps. If I were feeling well, I would join them—their laughter sounds like it starts in their toes. I show Carlos some of the biggest mushrooms I've ever photographed while we reminisce about the huge *Lepidella* from the hunt. When we get back on the highway, we all cram in together again. Once the going is smooth enough, I write a note to myself inspired by Carlos's table talk: "Always know the protagonists." I affix a small card with an old scientific illustration labeled "*Lactarius volemus*" that Diana gifted me to a page in my new journal. She notices and smiles, gives me a lollipop from her bag for my headache.

The crews of street cleaners we observed that morning are gone. According to our host, they hadn't belonged to any organization and were probably just normal people with free time who hated to see trash along streets that otherwise afforded stunning views. Through the one eye I can keep open, I watch as we pass distant mountains, black mineral streaks stark on limestone, like a record of rain tattooed down their backs, small towns at their feet. I duck my head down a bit to stop it spinning, and the matsutake scent wafts up at me from my tote bag. In this part of the world it's a late summer aroma, too.

If foragers are thieves, what of the matsutake in their mutualistic relationships with pine?

A few days before we joined up with this group, when we were based in Oaxaca City, we went to visit the hardest to hug (that is, the stoutest) tree in the world, the 1,500- to 3000-year-old Árbol del Tule. As the church bells tolled on the doubly sacred site (Zapotec and Catholic), the skies darkened to a slate gray, threatening to open up. Because I'd heard rumors that the tree is slowly dying due to intermittent drought, I ask the security guard how long he'd worked there (a while) and whether he had ever seen any mushrooms growing near the tree (he hadn't).[11] Minutes after I put this question to him, who should I see fruiting at the base of the nearby descendant of el Árbol but *Coprinus comatus*, which I would see for the second time that year, about a month later, near one of the oldest trees in my home state. If el Árbol is dying because of pollution and traffic and drought, it makes ironic sense that it is on this site where the two traditions battle.

Maybe foragers are only thieves in a system that is jealous of our ability to craft our own lives collaboratively. Jealousy is the engine of so much theft. Jealousy of one's high-femme middle school classmates, jealousy of another's land and culture. I am jealous of the pre-Columbian-ness of tejate. I am jealous that they can trace their unbroken lineage as people of the rain back so far. I am jealous of the abuelas at their ID tables. El Árbol makes me jealous. It's moments like these when I feel most United State-sian, when I want to bring it all home with me to enjoy. But if we're to live honorably, entangled, contaminated by one another, then let's allow that jealousy often shows us what we want, what we should then work to build out of the local materials we're still only borrowing. Not from a landlord, but from the Earth itself, which is so provident that someone had to invent theft in order to profit. Maybe what I actually want is the fact of an ancient heritage that cannot be buried or erased. So yes, I have to build or rebuild that, which starts with actively building the future here and now. If I'm jealous of trees and the lineages of others, then what I actually want is a history that is whole, to make a history together with you where our labor is playful and sweet, to step through this portal we've made of the present, toward an uncertain future, livable, real.

IN SEARCH OF *Schizophyllum commune* (THE CHARISMATIC MUSHROOM)

On sluttiness, purity, loving charisma, academic siloing, and getting it a little wrong to get it very right

Talking animals are for children and primitives. Their voices silent, we imagine wellbeing without them. We trample over them for our advancement. We forget that collaborative survival requires cross-species coordinations.

—Anna Tsing, *The Mushroom at the End of the World*

Everything can be explained to people, on the single condition that you want them to understand.

—Frantz Fanon, *The Wretched of the Earth*

Despite the fungal kingdom, queendom, theydom's reputation for housing a bunch of uncanny weirdlings, *Schizophyllum commune* has charisma. In this Western context, that feels revolutionary for an organism that deals in death and decay. The first time you pick up a stick with *Schizophyllum* fruiting from it and flip it over, you're likely to be struck by the strangeness of the hymenium, or fertile surface. I once heard someone describe it as crystalline—frost on a window pane. From a central depression stretch long, Y-shaped gills that appear split down the middle, giving it its common name, the split gill. The hymenium also looks a bit like warp-speed space travel (she said, adjusting her pocket protector), like the stars have elongated as you rush past them. Perhaps *S. commune*'s looks explain why their fruitings are so often chosen to play poster girl for "breaking" science stories.

A close-up of *Schizophyllum commune*'s wise and hirsute visage hovered above the headline "Mushrooms Communicate with Each Other Using Up to 50 'Words,' Scientist Claims." In April 2022, I was sent this and other pieces on the associated study nonstop, with messages appended like "You must have a direct line," "Ah, I get it now, you're a talking mushroom," and, simply, "!!!" Honestly, I was skeptical; like many taught in the West, I was raised to tamp down anthropomorphic impulses, to which these articles were almost certainly going to give in. At the same time, the hubbub around the study seemed like a lot of tired, scientistic goggling about something that feels intuitive (*Fungi communicate!*

Water is wet!), like an old story repackaged with buzzwords, one that frankly insulted fungal intelligence.

I'm bored by the continual discovery that the world around us really lives. At this rate, we'll be on the brink of extinction as a species, and the final ten humans huddled over a fire will observe nearby crows using bottle caps to sled down the windshield of the last, moss-covered car in the village, and wonder whether they're playing or not. (Say what you will about humans, at least we'll leave the crows a lot of bottle caps.)

I would paraphrase the conclusion of the study itself this way: The electric activity of some fungal hyphae has a structure that sort of looks like a language and if you take that as a given, which you needn't feel compelled to by the evidence, we've isolated fifty possible words in that language.[1] The researcher behind the study itself told *The Guardian*, "We do not know if there is a direct relationship between spiking patterns in fungi and human speech. Possibly not."[2]

Okay, a little anticlimactic. But believe it or not, I still enjoyed receiving this slew of messages from friends, acquaintances, former employers, and family. I was moved by how this outpouring spoke to broad engagement on a topic so obscure. The communication that I was most wowed by in this instance was not those fifty "words," but the fungal charisma (as in: *study me!*) that prompted the study that prompted the conversations we were having about the study's validity. But many stalwarts in the mycological community responded to the wave of media with simple dismissal. "Utter nonsense," said one oldhead on Facebook.

Perhaps it's because I'm not a scientist and perhaps it's because I'm not weighed down by a lifetime of fielding "utter nonsense," but something valuable gets lost, I think, in that simple dismissal. No matter what got the crowds there, it was energizing to see so many people from diverse corners of my contact list speculating about communities of life that pulse in soil and bits of dead wood around the world.

I loved how ready folks were to have their wonder reflexes tested by processes so invisible and humble. It feels right when masses of us rappel down from human scale: as when we christen a tardigrade "the water bear" and make much of their cuteness. When we weep at the latest list of tropical bird extinctions. When we make jokes about human dads because males in a newly observed species of fanged frog seem to be "hugging" their eggs.[3] The *Whoa, mushrooms talk, Maria?!* moment felt like love, and not just for me.

Science is a story a culture tells about the world based on information available to human senses. Varying degrees of rigor may be used in finding the story with the greatest truth value, but the investigator's notions of *what is* will always be a reflection of how they perceive the information they collect and how that information is collected. In Western contexts, scientists have often striven for universality, that is, the ability to make discoveries that apply in a vast array of settings. This sometimes works fine and sometimes is disastrous.

As a student of fungi who also teaches mushroom identification, I often see the havoc wrought by this impulse to apply a concept more broadly than it should be. So many of the fungi we talk most about in so-called Northeastern North America were originally described in Europe, which set the "type" species in their genera, that I often have to correct excited foray participants with old field guides. They might hold up a gilled mushroom with a clown-nose-red cap and call it *Russula emetica*. I'll have to explain: "Firstly, *Russula* fruits, or brittle gills, are basically impossible to tell apart without specialized knowledge and a microscope. But also, *Russula emetica* is not a North American taxon. It may not exist here." In fact, the accepted science frequently changes (along with the taxonomic names) when genetic sequencing is done on what turn out to be separate species that had looked identical. No kudos for guessing why I might be annoyed that organisms from this landmass are judged by European standards.

Linnaeus formalized the Latin binomial system for classifying species, which often allows for admirable precision. However, Linnaeus's attempt to apply his taxonomic scheme to *Homo sapiens* went (and perhaps you saw this coming) racistly. His type "species" was the European human, and he found the humans on other continents lacking by comparison. His descriptions of people by subspecies is a hoot to read up on, but I won't get into that here—we don't have the space and I can't not rant about it (I've tried). Suffice it to say that while Linnaean classifications are tidier than what came before, the world is too weird and too good at keeping secrets for any one way to describe everything.

Why am I roasting science when so many have rejected living their lives by its principles, putting us all in danger, you might ask? I'm emphatically not doing that, but am instead attempting to ease us into what at first appears to be a narrow gray area between binary options. In the binary, we either reject the truth wholesale, not vaccinating our children as a result . . . or we pretend science is an unimpeachable and unassailable set of facts that emerged, fully realized as such, from the realm of Platonic Forms. Instead I want to say, Yes, the vaccine will help make us safe as a collective, but that Tuskegee thing sure was fucked up, and *that* was science, too, so I get your suspicion, Black folks! The ideal may be objectivity, but scientists currently being trained must notice their unexamined assumptions and points of view, because science grows from multiple lineages, many of them recently or still tainted by eugenics, racism, sexism, classism, ableism (really, all the isms that attempt to segregate the truly human from the provisionally human). They must remember that even pure science is always *for* something, or someone. Meanwhile, that narrow gray area between the binary options, which looked so small and lightless initially, expands once we learn how to talk about it. The biological world can be our model here, with its dearth of binaries, with its sloppiness, with its defiance of our categories.[4]

Allow me to tread lightly here. I am not a biologist or an academic. The loftiest bona fide I will accept is that of "naturalist." I'm only a person who has volunteered, due to her enthusiasm for a world that is mysterious to so many, to act as a guide to earthy wonders as we currently know them. My job is to keep up with the science as best I can as someone who is self-taught, and to tell a compelling story about the organisms we come across and the organisms that *we are* in relation to those organisms. My goal, with any walk, lecture, workshop, or publication is to fan the flames of love and to show people their stake in their environment.

Once a man hired me to lead a birthday woods walk for his girlfriend because she became taken with all things mushroom after binge-watching a show about a *Cordyceps*-human zombie apocalypse. An outlandish science fiction series that painted an impossible picture of *Ophiocordyceps unilateralis* infecting humans got them out into the woods with me, where I could explain how long coevolution between a fungal pathogen and its host really takes. They learned—because they were drawn into an educational space by the fantastical, by the horror show of anthropomorphic mushrooms.

Now that I've trod lightly, allow me to shout this opinion: Scientists whose work has lessons for us are going to have to get a little sluttier! We need to take some risks, hazard failure, live pop-culturally sometimes, point to the charismatic if we are to survive! We're going to have to make bad jokes! We're going to have to white-rot the ivory tower down (sorry) and invite a more engaged public to dance on its ruin!

What I'm advocating is for us to get in touch with our inner "sluts" as an antidote to our terrifying and seemingly growing ability to compartmentalize and put hard walls between each problem. I'm not talking about the righteous boundaries that keep us from being taken advantage of, but the ugly ones that keep us from collective problem-solving and liberation: That typhoon is happening over there, not here, we don't have to worry about it. Those birds went extinct there, the drought is over there, the anthrax was loosed from glacier melt there. The bombs dropped on *their* children. What *she's* studying in the field doesn't affect *me* because, well, nobody has shown me how it applies to my life. Nobody's made it real.

When I say "slut," I am reclaiming the roguish, the messy, the scampish, and yes, the sexually free. I'm foregrounding the porousness of our bodies (the borrowed ones wrapped in skin, our institutions and careers, the landmasses to which we belong) and invoking our latent ability to become one with one another, making the beast with several backs. I'm calling for an ecumenical, polyamorous, miscegenating kind of science communication. Sure, I could have just said "interdisciplinary," but that wouldn't get at the need to use what god gave ya, so to speak, and to speak to who's around you, to better link disciplines and economies of attention that have become dangerously comfortable in their separateness. (Of course, literal asexuality doesn't exempt one from the charge to be epistemologically slutty—in my own life, ace folks have been among the least bound by convention.) The singer who happens to be a climate scientist as well, who writes a story song about riding his bike called "I Put the Fun Between My Legs" and performs it at an open mic, is exactly the kind of slut I want on our team. He's using the rizz for the greater good, and I hope to see that song hit number 1 on the Hot 100.

Schizophyllum commune is a bit of a slut. Or at least, each individual has great potential to be one, if it so chooses. You can of course be open to romance with any gender and choose celibacy. But it's the openness that could tip the scale toward finding a compatible match. A friend once sent me a Valentine's meme with a picture of *S. commune* that said, "Text from an Aquarius when they are flirting with you: This fungus has 20,000 genders and can mate with almost every individual in its species just by bumping into them." That is some seriously powerful charisma. Rather than responding to the Valentine with "It's actually closer to 23,000 unique sexes," I virtually blushed and said, "Aww shucks."

Schizophyllum commune fruits everywhere. I've scarcely been to a woodland where I haven't seen it or where I suspected it couldn't grow—it has been documented on every continent but Antarctica, has even been reported in the sinuses of a child and the lungs of a grown man in rare instances of mycosis.[5] I have delighted in fruitings smaller than my pinky nail's tip in the Northeast woods, and said wow in the Everglades at a specimen near a sunning alligator, with a fuzzed cap so robust it filled the center of my palm.

As *S. commune* moves, it relentlessly, recklessly mutates. According to a study testing the rate of its mutations, "the basidiomycete

Schizophyllum commune has the highest level of genetic polymorphism known among living organisms."[6] Does this mean that it's not the carefullest of beings, communing now, asking questions later? *S. commune* gets around physically as well as in the form of memes. I don't want to alarm you, but you are probably pretty near some of its mycelium now. A fuzzy, cute death-eater and death-dealer, it is "non-binary in the extreme," according to mycologist David Hibbett.[7] *S. commune* has what many would call a winning strategy.

We, on the other hand, have gotten ourselves into trouble as a species by buttoning up, being hands-off, isolating and then vivisecting, putting knowledge in ivory towers or discrediting it because it doesn't come in a rational enough package. But "rational" packaging has not led us to a rational place—we are daily committing slow-burn suicide, living as if the material world were infinite. Where does this come from but the accident of our dominant economic systems, which deny sex and death even as they're fueled by them? What would a true embrace of sex—and the concomitant queerness, cross-contamination, vulnerability, risk of being wrong—really, truly look like?

Egyptian doctor, writer, and feminist revolutionary Nawal El Saadawi gave an interview with Krishnan Guru-Murthy on the UK's Channel 4 in 2018, in which she delivered many incendiary and eloquent truths in a language not her first.[8] The words that I put up on the corkboard above my desk were these: "Education [as practiced in most modern societies] is based on splitting reality in compartments. They divide our head." Gesturing at different spots on her head, she added, "You put here religion and here gender and here medicine and here history. That's why we end by knowing nothing. But creativity is to unite. To link!"

I see a process of compartmentalization in how some identify in online mushroom spaces. As I've gotten to know more and more mushroom people—from lab scientists, to field identifiers, to poison control/toxicology personnel, to foragers and chefs—I've observed something strange happening. So many young women and femmes, whether they started learning about mushrooms for psilocybin, cuisine, or curiosity, eventually make a public disavowal of doing much collecting to eat. I fronted this way myself before snapping out of it, despite the fact that I was almost always eating, or at the very least sharing, my finds when they were plentiful and tasty.

Perhaps these women are simply growing tired of mushrooms as a category of food, or maybe they get solely into scientific inquiry, or maybe familiarity breeds a kind of contempt, and they've come to know mushrooms too well to find the idea of eating them exciting. While any or all of these could be true, I think there's also something less savory to blame for the decline in (public) culinary interest: the politics of purity. Compartmentalization. A separation between *us* true students of mycology who do not care about edibility—and *them*—those who are likely to make friends in this hobby by cooking delicious meals for people. Never mind that women have historically been important protagonists of the kind of ecological knowledge on which modern science is built. If you want to be taken seriously, an interest in the homely arts is contraindicated. Apparently nothing that nourishes, sustains, or is part of reproductive labor can coexist inside a mind taken up with the Serious Consideration™ of the organism.

My friend Blacki Migliozzi is a great example of not getting stuck in one box. He's part of the data journalism team at *The New York Times* that won a Pulitzer for their work visualizing COVID, and he was also a finalist for that prize for his work visualizing climate change—some stories he's worked on recently include: "America Is Using Up Its Groundwater Like There's No Tomorrow," "Monster Fracks Are Getting Far Bigger and Far Thirstier," and "The Illegal Airstrips Bringing Toxic Mining to Brazil's Indigenous Land." Blacki understands how a compelling narrative can draw the public eye toward urgent issues, as in "Where 2020's Record Heat Was Felt Most." (As a writer, I envy him for the same reason I envy people who make their living in film. Our culture is a visual one.)

Apart from being a laureled storyteller, Blacki is also a mycophile. He has the most unconventional mushroom origin story of anyone I know. He received his first magic mushrooms in utero, so in that way, fungi have been with him from the very beginning. Then he and his mother did them again, together, when he was a teen. It was a pivotal experience, one he now feels was undertaken to help heal various traumas from his early life. Unlike other "drugs," Blacki says, psilocybin was set apart as sacrosanct, something he didn't feel the urge to do to get high.

So fungi remained offstage for him until, as he was trying to figure out what he wanted to do with his life, he saw fractals on the cover of a book about mathematics and recognized the patterns from his formative trip. That was his sign to study math, which is how he met my older brother, a math professor, who, Blacki says, showed him that it's possible to put complex information into terms people could understand. Later, Blacki was minding his scholarly p's and q's while studying human-computer interaction at Georgia Tech and working toward his master's, when life again surprised him. A mushroom-obsessed friend of his named Liam, a charismatic food-sustainability activist who had big dreams of building an agricultural "makerspace" in Atlanta, died tragically on his motorcycle, hit by a drunk driver.

Before his untimely passing, Liam had applied for a Ford Foundation grant to realize the makerspace: a "unique integrated living system [that] will produce gourmet mushrooms, fish, and salad greens by upcycling organic waste biomass from local breweries, coffeeshops, and municipal arborists through a multi-stage bioconversion process." The grant was awarded posthumously. Friends, family, and colleagues of Liam's lobbied for the school to accept the money so they could move forward with his plans, and won. Blacki, who had volunteered to help in any way he could, ended up being the project's mushroom guy. It was a lot to commit to while he was also meant to be getting a master's, but Blacki is unfailingly resourceful. So he found a professor willing to advise him on "an amorphous mushroom master's project" and accepted the mushroom guy mantle thrust upon him.

Blacki learned what he needed to know about cultivation for the project by doing it. Liam had left a thumb drive with his vision and a fridge full of mushroom strains, so Blacki started setting up a lab armed with these tools . . . but without the benefit of ever having taken a biology class. Do you know what he did? Reader, he built a mushroom lab in the back of Georgia Tech's video game department. And he got so good at growing mushrooms that he was generating more biomass than was usable. He tied in computers by measuring mycelial run rates and contamination for the grows. He also branched out and started growing spirulina, a nutrient-rich alga, and used computers to measure its readiness for harvest.

Blacki really did the damn thing. He helped bring the work to literal fruition, honoring the memory of his friend and graduating in the process. He has continued to choose projects with charismatic main characters in his ongoing biological endeavors, from documenting life in the Superfund waters of the Gowanus Canal, to studying the world's cutest microbe (the tardigrade) and the lessons it can teach us about drought resistance. Looking at this snippet of his story, I'd like to challenge you: Can you spot all the instances of interdisciplinarity, of charisma, of academic sluttiness throughout?

By this point, most everyone has heard of and fallen in love with the Wood-Wide Web. People will often quote "facts" about it to me, proud to be conversant in this aspect of forest ecology. The history of this idea is a long one and I won't get into it at length, but here are some of the broad strokes: PhD student Suzanne Simard writes a dissertation about plants signaling to one another using the fungi that connect their roots in the '90s; major magazine publishes cover story about the research that went into it with the catchy title "The Wood-Wide Web" at around the same time so many around the world had started online lives; the idea of mother trees talking to and nourishing their offspring via fungi becomes a slow-burn fire that makes its way into television, books, movies, songs, and every other sort of media over the next three decades. If the culture this research touched were a spectrum, Eywa of the 2009 blockbuster science fiction epic *Avatar* might lie on one end of it, while Simard's own *Finding the Mother Tree*, a book of memoiristic science writing from 2021, might lie on the other. The public has taken the concept of the Wood-Wide Web and run with it as a metaphor for the connectivity and potential for cooperation of all living things. If the trees they had always imagined as discrete beings are actually talking and sharing and building community underground, then what else weren't they taught at school? For some, the Wood-Wide Web is the most compelling science they can remember learning—evidence against the "red in tooth and claw" rat race of constant competition they'd been told the world was.

The only problem with the Wood-Wide Web is that, well, it's unproven. Even that moniker is suspect in its tree-centricity, treating mycorrhizal fungi as a mere phone line for tree communication. As a mycophile I'm hurt by this vision of fungi; with every mischaracterization of them as tree accessories, I worry they'll fade back into the obscurity from which they've only recently emerged. Furthermore, in a paper entitled "Positive Citation Bias and Overinterpreted Results Lead to Misinformation on Common Mycorrhizal Networks in Forests," mycologists Justine Karst, Melanie D. Jones, and Jason D. Hoeksema examine the ways in which research into widespread common mycorrhizal networks (CMNs, composed of fungi linked to multiple plants) is ongoing, requires further substantiation, and has become dangerously distorted.[9] This misinformation, they argue, is now being transmitted to children who are learning about widespread CMNs in school. It's even reached discussions among non-mycologists about matters of forest management!

The gist of their paper is that we don't really know *all that*—we don't know that CMNs are in every forest, sharing food and signaling one another about danger. We don't know that CMNs operate between a mother tree and a baby tree, ensuring the success of seedlings. Some of the claims that go into this compelling popular narrative have never even been touched by the sanctifying light of peer-reviewed publication. Once you learn how partial and inconclusive the existing studies are, it's a little shocking how people have started to take the Wood-Wide Web for granted.

So, what are fungi actually doing with plants? Well, they *do* filter the spoils of photosynthesis from trees to myco-heterotrophic plants, like ghost pipes, that can't photosynthesize on their own and thus need fungi to live—which is sort of an incredible, inscrutable relationship in itself. More generally, as an article based on an interview with Karst explains, "the fungi draw nutrients and water from the soil and pass those to the trees, and also protect the roots from pathogens. And by using and storing carbon from the trees, the fungi also benefit the forest."[10] She adds, "Mycorrhizal fungi are essential for the growth and survival of trees, and have an important role in forest management and conservation practices, even if trees are not talking to each other through CMNs." It is also possible that some of the broader claims made about fungi-tree relationships will be borne out with further study; we just have to wait for those studies to be conducted in better conditions. And we have to accept that what we hear might not be universally applicable.

Justine Karst became a passionate debunker of the Wood-Wide Web story as it made its endless rounds on Twitter. I watched her tirelessly correct anyone who seemed to be unequivocally suggesting that CMNs were ubiquitous, or that any of the claims her review found to be tenuous or false were hard science. The cause was clearly a just one, yet something was nagging at me. That feeling became sharper when Gabriel Popkin, who wrote about the controversy for *The New York Times*, commented on his own blog that the WWW had "breached the walls of science."[11] Much as I appreciate his work as a science writer, his phrasing made me wonder: What kind of barriers *should* there be between science and the rest of us? Should there be walls like those around old cities and castles, with archers in the towers to make quick work of anyone attempting to approach without proper interpretive skills? Should there be moats?

I don't think so. I think there should be welcome mats for the unwashed masses (which one is always a drought away from joining, anyhow), and there should be stories ready for us when we arrive. People are going to spin yarns about the world whether they're sanctioned by replicated and cosigned capital-*S* Science or not—and scientists themselves, as we've seen, are not immune to the spinning of yarns based on emotions.

One corrective I've seen to the overzealous spread of the idea of the Wood-Wide Web came in a response to Karst et al.'s "Positive Citation Bias" by two scientists also versed in the literary method (which just means the study of what makes sentences sexy or writings rizzy). Their paper, centered on Simard's science memoir, is called "'I Struggled with Anthropomorphisms': On the Problem of Metaphors, Happiness, and Forests in *Finding the Mother Tree*."[12] In it, the authors suggest that while it may be too anthropocentric an analogy to say that CMNs confer something like "motherhood" upon a tree, perhaps a better metaphor *is* possible for these fascinating relationships. The authors don't supply one of their own, but as I read this, I immediately thought about Anna Tsing's cross-species entanglements and contaminations, and the metaphor that came to me was "chosen family."

Far from the heteronormative ideology underpinning the focus on a mother and her progeny, "chosen family" suggests collective adaptability, and reflects the ragtag messiness of both dynamic natural relationships and our imperfect, everchanging understanding of how these organisms make their way together. Rather than ascribing some brand of *Giving Tree*–type maternalistic self-sacrifice to the plant and her connections, and making fungi mere auxiliaries to this brand rather than sovereigns of their own biology, the notion that all involved choose these connections, based on their evolutionary needs at that moment, allows for organisms to be messy, communally wired opportunists—just as people are messy, communally wired opportunists.

As a smaller point, the authors also suggest that using cutthroat competition for resources as a lens (the way establishment scientists were often trained to view the give-and-take, push-and-pull of all natural processes) also has a metaphor at its center.

Competition is also a story we tell ourselves about the lives of other organisms, despite having incomplete pictures of the ways they eke out their existences. It's all a matter of translation for those who study the world. All of it, every crumb of life, every oak tree, every stitch of mycelium has ways of living and dying that the student must translate with human senses into human languages for other humans. It's just a matter of how skillful and how rigorous the translator is in carrying off communication in the new language. It is being versed in multiple languages and having heard a plethora of voices and having loved them all enough to take them seriously (that is, having a promiscuity of interests) that allows for the most communicative dexterity.

This kingdom has no shortage of enthralling, counterintuitive stories to tell, and we should tell them to the public every chance we get: a fungal parasite that commandeers an ant's body to propagate its own kind. A fungus in the tropics that makes webs to catch and eat leaves falling from trees. A wounded man in the Alps carrying a mushroom in which smolders the last fire he'll ever make. A fungus that mimics pollen to parasitize plants.[13] The amazingly stable throuple formed by an alga, fungus, and cyanobacteria that somehow takes the entirely new shape of a lichen.

Or consider this single sentence from a peer-reviewed study on carnivorous fungi like the oyster mushroom: "The nematode-trapping fungi produce several volatile compounds that mimic nematode food and sex cues to lure their nematode prey, eavesdropping on nematode ascaroside pheromones, and they develop various nematode-trapping devices such as adhesive networks, adhesive knobs, and constricting rings to capture the prey."[14] I would watch that A24 horror movie!

Of course we should fight misinformation and the kind of storification that erases all nuance. But purity doesn't exist. Purity will never exist. Purity shouldn't exist. Science needs art and vice versa—they always have and always will. (Remember that Johannes Kepler regarded his third law of planetary motion as a harmonic law, expressing the music of the spheres.) Remaining cloistered in our halls of science angrily being right and scorning the popular story, the one that runs away with an idea, will do us little good if we aren't also trying to write a popular story.

Remember, kids, we can't even *not* anthropomorphize the stars. This is how we've always been. It's only recently that some of us decided it would be better if animal, vegetable, and mineral lacked spirit, and we did it in order to factory farm and frack with aplomb.[15] We look to the sky and see dots of light and project twins and people moving water around and shit. That tardigrade was always going to turn into a bear.

So, is there such a thing as informed dreaming? What does a responsible speculative space look like? "Do whales judge us?" I think I have found tentative answers to all three questions in the work of Bathsheba Demuth, author of the ecological history *Floating Coast,* who asked that last, provocative question about whales in a talk she gave a few years ago, a question that came from her ties to an indigenous community she had lived with for years.[16] The Yup'ik hunt whales traditionally, and they understand the whales as sometimes offering themselves up to the harpoon if they find the hunter respectful enough. After a lecture that had me dreaming, the Q and A session was electric with "Wait, are you serious?" and "You can't possibly mean. . . ." In talking with my partner about the fallout from Dr. Demuth's ideas, I wondered why her interrogators found it so difficult to stay in the speculative space she asked us to inhabit for about an hour of our long, Western lives. He wondered what's so dangerous about that space. That space where another intelligent species, whose lifespan is so much longer than ours that we aren't even sure what "old" means to them, is able to simply approve or disapprove of a human creature's actions.

In her talk, through stories of Yup'ik, US, and Soviet whaling, Demuth highlighted that while cycles of life and death are materially inextricable from the way we consume resources, those cycles were absent from both US and Soviet moral/economic cosmologies, their concepts of what should make the future. A culture cannot sanely ignore death, yet this denial drives what so many call progress. The desire to ignore death is one ingredient in mycophobia. But to ignore death is to cage sex.

Carl Safina, like Bathsheba Demuth, has courted controversy by issuing provocations meant to shift the ways we relate to the more-than-human; specifically, Safina wishes to question why anthropomorphism remains so utterly verboten as a springboard for inquiry. In a blog post, he points out that those turning the study of animal behavior into a science—that is, operationalizing it so others might take it seriously, proving it could be objective work while brain science was still in its infancy—did so by defining that science against folklore and mythology, which sought to explain the unique interiorities of other animals by simply projecting our own cognitive and emotional baggage onto them. But he suggests we've gone too far in the other direction, to the point where certain experimental questions can cost people their jobs. He also makes the excellent point that "it" is not a way that field scientists refer to the objects of their study.[17] That language comes to us straight from the lab, on a sterile tray. From the least slutty place possible.

I'm in awe of those accomplished science communicators who move through the internet casually teaching people, like comic artist and author Rosemary Mosco, writer and podcaster Alie Ward, "Black Forager" and America's sweetheart Alexis Nikole Nelson. I'm equally in awe of the scientists who came out of their institutions of higher learning primed to bring their research to the rest of us in engaging ways: "Hood Naturalist" and ornithologist Corina Newsome, mycologists Patricia Kaishian, David Hibbett, and Merlin Sheldrake, biologist Gordon Walker, and physicist Chanda Prescod-Weinstein, to name just a few. Rounding out that list is the luminous Robin Wall Kimmerer, who is inspiring a generation of scientists, naturalists, and others who love the world around them enough to question the ways they've been taught to take it in. When I lead a mushroom walk, I hold these words from *Braiding Sweetgrass* in my mind so as not to lose the way: "When I am in the woods with my students, teaching them the gifts of plants and how to call them by name, I try to be mindful of my language, to be bilingual between the lexicon of science and the grammar of animacy."[18]

During a talk over Zoom, poet and essayist Anne Boyer gave me a singular gift. When I told her I wasn't interested in mushrooms as a metaphor (I wasn't even 100 percent sure what I meant by this; perhaps just that I needed to ensure writing about them was an act of magnification rather than reduction), she said, "We are the metaphors for them."

With those words, it felt like she pulled me inside out and diagrammed all my bodily miracles by their humblest names. My mushrooms and I set up an altar to our Bacterial and Archaean progenitors. In a sentence, she convened a quorum of my non-human ancestors and later, when I picked the words apart in a notebook, I overheard these fellow Eukaryotes speaking about an Earth in which the first mammals, tiny, shrew-like creatures, were a far-off glimmer. I try to internalize the humility and wonder inherent in this simple phrase every day—and while I don't always succeed, every day of trying is a victory.

With sluttiness as the spoils of the hard work it takes to remain open to the world and its people; with skepticism born of habits of critical thinking for which we can thank the multitude of teachers in our lives; with a long view taken from the knowledge that we are the metaphors for them; and with a compass built from the salt, gristle, sticks, bone, and blood of the need for justice, we can create collective ways of knowing that value a planet in balance rather than one savaged by the profit drive. Worldviews are pliable, remember: We are the metaphors for them.

IN SEARCH OF

Undescribed Species

(ALL WE DON'T KNOW)

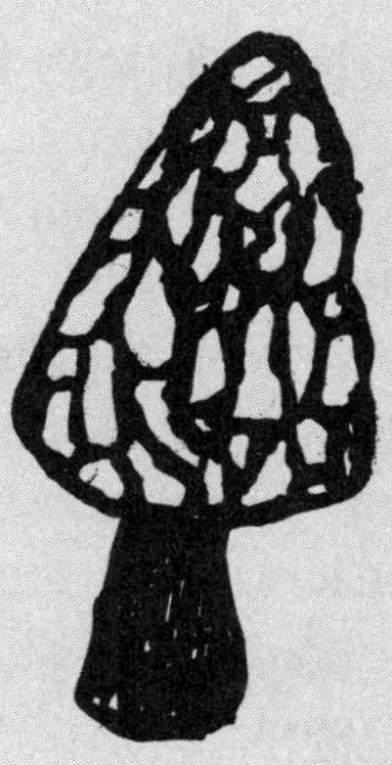

On the names of things
and why they matter, true
community, being willing to get
filthy, who needs the future?,
death and indeterminacy,
learning from the young elders,
the lights in the forest

Betty Isaacs told me that when she was in New Zealand she was informed that none of the mushrooms growing wild there was poisonous.* So one day when she noticed a hillside covered with fungi, she gathered a lot and made catsup. When she finished the catsup, she tasted it and it was awful. Nevertheless, she bottled it and put it up on a high shelf. A year later she was house cleaning and discovered the catsup, which she had forgotten about. She was on the point of throwing it away. But before doing this, she tasted it. It had changed color. Originally a dirty gray, it had become black, and, as she told me, it was divine, improving the flavor of everything it touched.

—John Cage, composer and founder of the New York Mycological Society

"Tusk" is the word my mind supplies when I see the arc of bone reaching above the tall grass in a field in rural Northern Virginia. Geese overhead laugh in their V-formation as I head for the flash of white amid green. The sun is low, covering all in long, gilt shadow. "Tusk," but the word I want is "antler," and there are two of them up ahead, attached to an almost perfectly intact skull. "Antler," but I understand why "tusk" should have rushed to the surface of my citified consciousness the way it did. My fascination is mammoth. I feel an archaeological thrill to be happening upon a fellow creature's whole scaffolding like this, without the stench of decay or flies, without the bloating that occurs when a death is of the summer and is yet unloved by a full range of natural processes. It's not often that I see the remains of a deer that likely died of natural causes. There were plenty of deer and other animal bones strewn about the property, but this is the first full skeleton I've seen.

The ribs are adjacent, stark empty and sun bleached, their visual meaning so different unenclosed. Tattered hooves are still attached to the creature's long legs. The skull comes free with a bit

* NB: THIS IS NOT TRUE.

of coaxing from the brambles that had begun to snarl it back to the earth. The lower jaw is attached only by fur baked by the animal's own fat into hardness—it comes away when firmly pulled. The whole architecture of what is left of his head is so intimate. Here, I can see the whites and grays and blacks and browns that make up this white-tailed deer's winter coat. I can see his fantastically long eyelashes, the black and brown decay of his beigeish teeth, the dead tick still embedded in his fur. His skull is only half clad—the left hemisphere more so than the right, the flesh pulled up over some of the absence of an eye on one side, as if the former deer wears a low-slung hat that dips below his brow. On one side the preorbital vacuity is just covered by well-furred flesh, on the other that edifice is exposed. He has an eight-point rack: that is, eight spikes of bone with which to mark a tree with his essence or warn a competitor off his territory. This is the most metal thing I've ever encountered. It reminds me of the strange mutant bear in the film *Annihilation*—the one that records the voices of people it eats, its skull half exposed. My deer skull is repulsive and vaguely terrifying, and something in it pulls at me. I have to take it with me. The brief hesitation about possibly leaving it for squirrels, voles, and other creatures who depend on the nutrients provided by animal armature is fleeting—maybe the skull can nourish me, somehow?

It is the beginning of morel season and I'd hopped on my bike destined for the old-growth woods on the property—I wanted to investigate the terrain at the bases of the elder tulip poplar trees and maybe snag some fresh "dry land fish" (morels) for dinner. It is good I have the Moroccan market basket I use for bicycle foraging and a sweatshirt to change into—now I carefully line my basket with the shirt I'd been wearing, not wanting to contaminate the basket's interior with any leftover yuck lingering despite how cured the flesh seems.

It is another windy day in Upperville. Fires have lately hazed the air. The grasslands native to this part of the region, between mountains and forest, bow and flash with each change in the wind's direction. I imagine my deer in life, how his winter coat might have come off in a wind similar to this one as his reddish coat grew in for a summer he would not live to see. In the past, when I'd observed deer shedding in that way, I'd thought it was due to some illness or the stresses of living in the city. Pure projection. But my deer would emerge from that mottled, mangy-looking liminal stage with fresh threads readymade for warmer weather. Again, pure projection to imagine a living future for a creature I am already beginning to think of as kin, though his skull has only freighted my walk for about half an hour. A friend texts me as I head back to where I am staying, morels forgotten, two antlers sticking up from the top of the pack on my back. He says he misses me and wishes that just now, when he'd passed woods we used to frequent together, I'd been in the car to demand he pull over so I could investigate something potentially fungal that I'd seen beside the road.

"Not much to spot yet up there," I text back—he is in Massachusetts where it is still functionally winter. I send a photo of the skull in the grass bathed in pre-sunset's beneficence. In the photo, it is propped up against my basket.

"Holy shit, did you take it?" An unnecessary question—this man knows me well. After a grilling about how I planned to dismantle it (the short answer: "I'll figure it out"), he expresses his disgust and says, "Cool souvenir, though."

Souvenir. Is that what I'm in the process of making? A memento? Of what? Neither word seems sufficient, because the life powered by the gray matter that used to occupy this skull is top of my own mind when I look at it. Could the skull be reduced to a visual shortcut to memories of this artist's residency, for instance, or of my time getting to know this landscape? Or would my sense of the spirit of the deer blaze up just as often—the life I imagined for him, both in his past and in the future that never obtained for him, here in the Virginia countryside? Virginia. That could be my skull's name—Virginia to queer it up despite the antlers. "Verginius" is what I've just learned is the masculine form. The Latin name for white-tailed deer is *Odocoileus virginianus*—the first, or genus, name meaning "hollow teeth." Those of my specimen rattle around in his skull, maraca-like and full of holes, as if he's starring in one of those nightmares I'm always having about structural issues with my own head of teeth, the dreams Jung said signify rebirth.

The white-tailed deer is in the fossil record before the ice ages and appears to have been officially described by Western science in a joint effort between three Europeans, centuries apart. In 1588, Thomas Hariot, one of Sir Walter Raleigh's main coconspirators in his attempts to colonize Roanoke Island and "discover" "new worlds for gold, praise, and Glory," wrote a book toward that end.[1] In *A Briefe and True Report of the New Found Land of Virginia*, he makes note of deer that "differ from ours onely in this, their tailes are longer and the snags of their hornes looke backward." In 1777, the thoroughly named Eberhard August Wilhelm von Zimmermann published *Specimen Zoologiae Geographicae Quadrupedum* to establish the geographic ranges of the world's mammals. He apparently did this for shade reasons, which I love, putting together a map of mammalian species distribution around the globe to refute the commonly held and biblically founded notion of the radiation of all animals from one central location. Three years later, in *Geographische Geschichte des Menschen und der allgemein verbreiteten vierfüssigen Thiere*, he properly proposed the white-tailed deer as *Dama virginiana*, favoring them with the genus name of the European fallow deer.[2] US colonists also called them *Dama* at the time, despite the genetic distance between fallow deer and white-tailed deer. The *Odocoileus* bit comes from a multi-hyphenate, Turkish-born Frenchman named Constantine Rafinesque, whom Audubon, the scoundrel, called crazy for being too obsessed with "the natural world," for perhaps wanting too badly to name a lot of species.[3] He named thousands. In 1832 Rafinesque wrote in the publication *The Atlantic Journal* that he'd been given teeth ripped from a jawbone in a cave in Pennsylvania, of an animal he could not identify, which he decided belonged to a perhaps undescribed species "near to the tribe of goats or dwarfish oxen." Scientists later realized the teeth he described were those of white-tailed deer, and codified the genus name *Odocoileus*.

But in 1948, US mammalogist Philip Hershkovitz pointed out that, wait just a minute, our boy Zimmermann had gotten there first.[4] There was brief chaos, with some scientists using *Dama virginiana* and others *Odocoileus virginiana* to refer to the same animal (can you imagine?!) until the International Commission on Zoological Nomenclature expressed Opinion 581 in 1960 (which I also love—bit of a late start, but I may start to number my own opinions). It acknowledged Hershkovitz's *well, actually* and made clear that he was right, but decreed that *Odocoileus* was what we were gonna continue to say, moving forward. If mushroom people have taught me anything, it's that those who study deer are probably still doing battle regarding cervid taxonomy. Looking into that is not something I can afford to do right now, though the beautiful white expanse of the search engine box is calling me.

From three to five million years ago to 1960 CE. That's a long time to go without a real name to sink one's hollow teeth into, right? Of course, these animals didn't go all that time unheralded by *Homo sapiens*, they just went all that time without being accounted for just so, using the Linnaean system of Latin binomial nomenclature, by a person, preferably a man, of European extraction, in a publication that wouldn't be ignored by the continually emerging Western Scientific Orthodoxy. In my encounters with historic mycologists, ecologists, historians, and botanists over the past several years, I've found myself screeching "Why isn't she a household name?!" and "Why does nobody know that they were doing that centuries before the people who take all the credit?" too often, and I am too Black a woman to believe that even a righteous fraction of people who made life what it is today have gotten their props. But the Wampanoag people, whose stolen land my apartment rests on, may have called the deer *Ahtuhqah* as they communally hunted them for millennia over the course of countless autumns and winters.[5] The Wampanoag who still speak that language may still call them that. The Manahoac, who came into contact with the first Europeans bent on settling Northern Virginia, where I found Virginia the Skull, may have called the deer *Witāi* as they burned the landscape to ensure the species diversity in the grassland browse that their quarry favored.[6] The

Manahoac who still speak that language may still call them that. I don't presume to know what white-tailed deer call themselves.

It's lucky my housemate has had to leave the residency early, though I miss her—this means I can go about the process of denuding the skull without bothering anyone else. There is no pot in the house big enough to boil it, which is a relief, even though the inability to introduce fire's intensity will make the problem more intricate, laborious, and messier to solve. I would have had serious misgivings about sending that kind of vapor into a working kitchen (as the antlers would preclude putting the lid on). Armed with kitchen scissors, an electric kettle, two buckets, a toothbrush, bleach, a knife, and determination, I peel the skull in fits and starts over the course of three days. Sometimes, I must step away lest my gag reflex make a further mess of the work—other times I come to an immovable section of flesh that needs further rendering and cannot continue. Virginia's rancid fat smells like a bag of shellfish carapace in the July sun. I have no gloves, so when once I must tug a tuft of fur away from a hard-to-reach spot under an antler, and some of the boiling-water-softened meat gets under my fingernail, I almost give up, stepping back from myself, wondering what this is really all about. What if I've truly gone screwy about the faculties during these isolated five weeks, and this or something like it was always going to be the outcome? Breaking down this dead thing that nobody asked me to break down . . . in fact, a few seemed to think it much too gross to take on.

But I wash my hands until they're tender, square my shoulders, and look Virginia in his lack of eyes—this has turned from a wisp of mystic intention to some sort of sacred charge, a needful ritual, a way to honor something (what?) that, if abandoned, might turn into a monster. This feels like a harvest too costly to waste. So I return to work. About midway through, when the skull looks as if it has a wet, mushroom-cut-shaped toupee, I stop at length to admire the black river I've uncovered running down the front of the skull, the way it winds against the newly exposed off-white and light-brown bone, its zigzagged meanderings forming one cursive line between the eyes with two tributaries that link between the antlers. Having uncovered some of it means I must continue, expose the rest. I press on through flesh and fur. There's a bruise gathering at the bottom segment of my right thumb from plying these dull-ass scissors. I should be packing or writing or reading. Instead, I'm wondering how I would be different if my family had stayed rural, kept the farm.

The cursive, or sutures, between Virginia's eye sockets are a way that those of us with skulls accommodate growth—one way our interior hardness is actually a practice of softness, of flexibility. At home, weeks from now, I'll examine Virginia when I need reminding that there are few ends that are true conclusions. What we think of as death, as *that's all, folks,* is so often transformation. My grandmother appends "if life spare" to the end of any talk of the future and has been doing so ever since she reached her three score and ten, three decades ago. When a book with one's grandmother in it molders in the basement of a crumbling house, and eventually that book turns to soil, and from that soil emerges brisk creation, perhaps it is exactly the life that was spared, and everything it wasn't that fell away. I trace the suture-squiggle's blackness with my finger—it reminds me of spalting in wood, that fungal turf war made visible. A tree's end stamped with the surest sign of living, the struggle. Fungi finger their way into everything. I can see them up ahead, breaking it all down, building it back up again.

Listen, we could argue till you fray your vocal cords about whether Carl Linnaeus was a dirtbag (he was), one of us screaming about how his system of classification was used for evil (it was), but the fact of the matter is, the whole enterprise has had some pretty far-reaching uses, scientifically speaking. The first woman with both the privilege and formal training to officially use the Linnaean system to describe fungal taxa appears to be German botanist and mycologist Catharina Dörrien. In her 1777 publication *Directory and Description of All Plants Growing Wild in the Principality of Orange Nassau,* she named two lichen after herself—*Lichen centrifugus var. major Doerr.*, and *L. centrifugus var. minor Doerr.*[7]

Though Sigrid Jakob recently joined that same badass lineage of "amateur" mycologists to name a species, the German-born president of the New York Mycological Society (NYMS) did not name the goose dung fungus she described from a park in the Bronx after herself. She and her two coauthors agreed upon the "tela" in *Sporormiella tela* for beautifully etymological reasons. When I heard she was in the process of describing a new species months before the paper came out, my delight for her went beyond the pleasure in seeing tenacity rewarded—it felt like a team member had medaled. It makes the win all the sweeter that she medaled in something as earthy as noticing a fungus whose job is to recycle the nutrients in dung. As we chat via teleconference in April, it's late afternoon for me and early morning for her. She's in Aotearoa, or New Zealand, for three weeks with a group of eight other mycophiles she calls her coconspirators, with whom she's "documenting the hell out of everything—ferns and liverworts and flowers and trees . . . trying to understand the whole ecosystem. It's really just permission to completely nerd out on all the crazy, amazing beauty all around us."

Probably like many around the world who have never been, I associate Aotearoa with the stunning lushness of the Shire and the bunkers of billionaires. What she's doing sounds like paradise. I make a note to check this team's iNaturalist and social media accounts for photographs of what they're seeing. I imagine the pictures will go pretty hard when I do, as Sigrid herself earned an MFA in photography and her coconspirators include luminaries in the world of fungal image capture: Stephen Axford, Alison Pollack, and Christian Schwarz.

Sigrid was the perfect choice as a member of this motley crew of mostly "amateur" mycologists. The fact that she ushered a new-to-science organism into the official record was not happenstance, but the result of having set that goal for herself, of having an enviably keen eye, of the assiduous study of the careers of female mycologists, of implacability, and of a willingness to dig into the dirt of life, in the oldest sense of the word "dirt"—that is, excrement.

Anyone who knows Sigrid knows she's a lover of dung fungi. She's one of two people who come to mind when I think of those preternaturally unbothered organisms, and if my own text messages are this full of shit—that is, people wanting to know "What's this thing growing out of horse manure?"—I can only imagine what the text messages of a person who has an Instagram account dedicated to dung fungi look like. When it's freezing out, up here in the Northeast, we know what Sigrid is doing while so much of the rest of the mycelium is in winter repose—she's looking for deer scat. One January day, she had collected several containers of the stuff on the edges of a giant field in Van Cortlandt Park where the ruminants have their latrine. As she headed to the subway to take her haul back to her home for incubation, she attempted to avoid stepping in the leavings of the hundreds of geese who now permanently live in the park. Sigrid had not had much luck with goose poop in the past; it turns green and then black and is "otherwise unappetizing," in her words, so she hadn't been in a hurry to return to it as a substrate. But in that moment, with the more promising deer shit in hand, and more winter ahead of her, she said, *Why not, why shouldn't I give it a shot?* The decision was made to give goose shit one last try. Then what?

"Then it was just laying around for six weeks doing absolutely nothing," she tells me. When Sigrid says it was laying around, she means that dung takes up a couple of square feet on her work desk in her marital bedroom, housed in Chinese food takeout containers—and this, dear reader, is the image she shared that made me want to be her friend forever. She jokes that her husband is a tolerant man, but we know that he is a lucky man.

At the six-week mark, when she was ready to throw it out, this last chance piece of goose shit redeemed itself. She noticed there was a black crust on its surface, a sort of weft. Unusual. When she peeled it back—like ya do—many little orbs were revealed. This was very unexpected indeed.

"So," she said to herself, taken aback, "this fungus made a little carpet and was hiding these little eggs under the carpet? What the hell?" When she tells me this, I think, *That just sounds like normal ADHD housekeeping*, but I keep this to myself.

After some microscopy, she could tell it was *Sporormiella* (her favorite genus), and when she couldn't find anything in the literature to help her toward the species, she posted about it to a German dung fungi forum replete with the kinds of nerds who could get wind of that weft and tell her without skipping a beat that she should talk to Mike Richardson, who found something similar twenty-five years ago. So she went looking and found him on ResearchGate. His profile said he was eighty-five and essentially retired from everything except coprophilous (poop-loving) fungi, which, of course, was perfect for Sigrid's purposes. The former professor had in fact found the same thing in Iceland twenty-five years ago, and he was pleased as punch for her to take the reins as lead author on the paper he offered to write with her. He was not big on genetic sequencing, while that's a specialty of Sigrid's, but was adept at the parts of the writing process that required scientific jargon, for which Sigrid was grateful. She reckons they split the considerable work fifty-fifty.

After some back and forth, and the addition of a third author who gave the whole affair taxonomic clarity, *Sporormiella tela* was described a year after Sigrid collected it, debuting in the *Northeastern Naturalist* next to articles about snake assemblages in coastal Maine and three different shore birds on Fire Island hatching the eggs of a different bird species.

The exciting part is getting to name it, right? "Tela" means loom in Greek, and refers to the woven, carpet-like structure the mushroom makes for itself, this weft acting as a sort of protective layer. As mentioned, it had been found in Iceland, and its genetic sequence matched ones taken from soil in Tibet and Kashmir, places of high elevation and latitude—cold places. Sigrid wonders if this adaptation is the fungus weaving itself a little coat. This weave is distinct, and thus, the name. Sigrid believes (and is correct) that if we're using binomial nomenclature it should be descriptive, rather than shouting out a dead person.

Sigrid was born into a large family in a small village in southwestern Germany. She recalls when she was around six, walking through a forest littered with vibrant, tentacular creatures, like still, red octopodes, but on land. She felt like she'd stepped into the pages of a fairytale. "I didn't talk to anyone about it, but it left a really deep impression. Ever since then I've been fascinated by the beauty of fungi." She also remembers family trips that ended in the collection of buckets of horse mushrooms, which she characterizes as a holdover from the rediscovery of foraging during postwar austerity. As she says this, I'm struck by the differences and continuities between her solo walk as a young child, the family foraging expeditions, her solo treks in New York City in search of dung, and the trip she's currently on in Aotearoa—all with their different levels of play, labor, and sheer fascination.

Time and again I hear it from mushroom people, and time and again I've said it myself—finding the other mycophiles was the first time in my adult life I felt squadded up.

"Being part of a gang" is what Sigrid says when I ask her what keeps her going out to parks on the weekends. "I've never really been part of one, or a movement, or anything like that—I've always been a bit of an outsider, and to discover this sort of global sisterhood or brotherhood—suddenly there's a place to belong." It makes sense: The person who prioritizes walking achingly slow, at mushroom speed, through the forest, whose idea of a good time is rolling around in the duff, turning over logs, and looking closely at the intricately patterned surface of a fungal crust with a jeweler's loupe is a particular kind of person. Where else are you going to find folks who want to know you because of, not in spite of, the fact that you collect feces in tubs on the weekend? Don't answer that.

When Sigrid became president of NYMS, she knew she wanted to make the club more welcoming and more adept at finding The Others, no matter their background. She reasoned that if she had been intimidated to join the club as someone with her education and other privileges, it must appear especially forbidding to someone without those advantages. But when she looked at the stats—how the club visits twenty-two parks around the city every weekend, yet before Sigrid's tenure they were hardly ever joined by people from the Bronx or Staten Island, for instance—it was undeniable club outreach wasn't all it could be. But it's gotten better. Sigrid knows it's just a start, what they've accomplished with newer partnerships with organizations like the James Baldwin Outdoor Learning Center, the Bronx River Alliance, and Latino Outdoors. But the work of being in true community, especially with people who have been actively excluded from outdoor spaces, is unending. It's the main reason an introvert who doesn't love organizing found herself in the role of president. If the work is unending, then a solid start is key.

Gangs and squads can be hard to come by. In the US, houses of worship used to be a primary third space, but fewer people than ever claim a religious affiliation.[8] But in the pursuit of mushrooms, and with the resurgence of community naturalism more generally (here I mean documenting the other organisms we inhabit the earth with, not getting buck nekkid in public places; but I support both with all the necessary caveats about consent), the third space is free, there is no faith requirement, and it endows those who take to it with a stake in the land. Over the years, I've watched mycophiles coconspire across divides of race, age, ethnicity, educational background, disability status, class, gender identity—it's one of the most heady aspects of the hobby. Despite the differences in where one might come from, how likely is it that the only things you have in common with your neighbor at the foray is that you're both circling an Eastern Hemlock, have your pants tucked into your socks, and reek of citronella? You may both volunteer for the street clean-up crew. Or you recently lost your mothers. Or attend meetings of the herbalism society. Or you're both neurodivergent. Or contribute to the community fridge. Or you both just failed the chemistry final but love using KOH in the field. Or you're guerrilla gardeners. Or you both suffered horrible breakups. Or maybe you're members of a radical environmental protection cell about to risk arrest by camping at the top of a tree to block the building of Cop City. Or you're both members of the same affinity group, whose goal next week is to block a shipment of missiles but whose ultimate goal is to dismantle the US, brick by brick, plank by plank, decomposing it from within like the fungi you study.

But maybe I'm only describing a dream.

Another of Sigrid's coconspirators, her colleague at NYMS, and the brains, spirit, and network behind their annual Fungus Festival is artist Sneha Ganguly, also known as Kali Mushrooms. Kali is the goddess of death and rebirth, whose Sanskrit name, Sneha informs me, is etymologically bound up with blackness, death, time, and ink. In Sneha's imaginary, mushrooms became a manifestation of Kali. Sneha is also an art handler, preparing, installing, and transporting artwork to ready it for showing. In the course of doing her job, she quickly became acquainted with aspects of the art world that were less than savory: its tendency toward overconsumption and wastefulness, its lust for hoarding away luxury. In asking herself how to belong in this space when those things are anathema to her values, she heard the answering call of mushrooms and of biological art—art that has a life cycle and can't easily be consumed or bought, art that will decompose. Sneha used to live in an arts co-op in Bed-Stuy. She values communal art-making spaces, and toward that end facilitated a workshop called "Surviving the Anthropocene With Fungi," in which participants, among other activities, made medicinal tea using fungi, then made paper from the pulp. Mycophiles seem to love sharing the practical skills our study of mushrooms has imparted, simply as an outgrowth of being in community with both fungi and our peers. It seems a natural outgrowth of that community to orient these skills toward forming new traditions that will usher us into a sustainable future. (We'll see later how Sneha's practice bred a mycofuturistic collaboration with an eight-year-old child, in the form of one of the most beautiful works of art I've ever seen.)

Anyone I've talked to about my teaching in the past couple of years knows that I've become a little obsessed with what feels, to me anyway, like a unique pedagogical problem. Back in June of

2019, I proposed running a class for teens at the creative-writing nonprofit where I teach entitled "Whose Future? Our Future!" The plan was for the high-schoolers to busily write against the dystopian grain that seems to have a complete stranglehold on all the stories we tell about what's ahead of us on this (and even other!) planets. I planned to set these kids this task while I cackled in my supposition that this could be a practical first step toward a more livable world. You can't be it if you can't see it!

The gist was this: We would read, watch, and listen to Afrofuturistic and solarpunk texts, then talk about how the futures in these stories were and were not sustainable, livable, the kinds of eventualities we wanted to inhabit. My prompt asked them to include a relic of our current world as an anchor for their wildly imagined future world, and the only other rule was? No dystopia! None.

The class, which I've now taught several times and will continue to teach as long as someone gives me a venue to teach it in, starts with students naming all the media they can that deals in the future. I ask that they keep the titles earthbound. I make a list of these titles on the board. Predictably, they are mostly dystopias. Some exceptions are *Star Trek* (not earthbound) and a manga whose title I can't recall. This exercise has never not proven the point that imagining a time when humanity has its shit together is not that easy. Or, people are imagining these stories all the time, and they don't tend to get chosen to be made into blockbusters the way fascist futures in which our children are cannibalizing themselves do. In any event, the next step is to talk about why. The suggestion that a good future would be boring is dealt with—conflict does not only arise, even in stories that take place in the present, from universal oppression or more people than necessary dying in climate catastrophes. The idea that the way things are currently makes a livable future impossible? Grim, but fair. The list of reasons it's hard is extensive. Generating it feels cathartic. Like we were talking about something more than storytelling.

Then I play them *Earth Mother, Sky Father: 2030*, a short film by Kordae Henry. In it, a voiceover explains that, in 2030, Afrika's rare earth minerals are no longer being shipped to China to make technology for the world, but that Afrikans are "refining raw metals on Afrikan land and building wealth from the ground." What follows is a stunning "ceremony to the God of rare earth." A dancer uses their body to operate machines that harvest from the land. The material never gets named, but the process is quite the literal dust-up. The dancer's moves are jerky, spasmodic, and fluid in equal measure. They are the machine, and the machine is them. At one point in the ceremony, they don a ceremonial mask that looks like what happens when a heap of disorganized jewelry gets all tangled together. It mirrors the Cthulhu-like profile of one of the earth movers they operate—a machine whose parts fly together in time to the ritual movements of the dance. In the background are products of this mining that look like pyramids. When the ceremony is over, the earth material goes back into the Earth.

During a recent talk offered by Boston Ujima Project on the topic of fugitivity, poet and theorist Fred Moten was asked how the experiment of Black life in the present shapes the conditions of tomorrow. Moten responded that we live within the nightmare of Eurofuturity. That we should call into question the desire to determine the future. That working with the present seeds the future. That perhaps we should develop the ability to detach ourselves from the dream of being able to determine the future.

Initially, I felt called out by this wisdom. What was my obsession with trying to find seers and prophets among my students and peers? Was it a desire to compete with the enemy's vision, with their organization, with their utter and absolute determinacy? To have something plotted and ready when the argument no doubt arises, as it always does, cynically, *Well, with what will you replace this dystopia that is working for me and mine?*

I'm trying to remember that one way to be a good metaphor for them is in our indeterminacy. The movements that continue to fruit year after year are the ones that myceliate their way into the sources of sustenance and adapt to their environmental conditions—going dormant when necessary, moving in the shadows, blooming in all their vividness and numbers and strength, in all their sluttiness and flexibility and diversity, through concrete and soft earth, at the tops of trees to dare them, on the forest floor, in the campus quad, through the burnt-out frames of cop cars, clashing but bending, sinuous, in the edgelands, on the commonwealth we've always foraged, on rematriated land, everywhere despite their fungicide, fearless, sleepless, deathless.

I took care not to define this new word, or to provide parameters as to what it signified, but I asked most of my interviewees what the "mycofuture" meant to them. Almost everyone either took a dim view of anything we might call a future or had all the political misgivings that stood in the way of any far-reaching livability. Some interpreted "mycofuturism" as a continuation of the current commercial "movement" toward mushrooms, which has slathered "myco" on every label, with the regular players, the ones who often win at capitalism, benefiting the most from the continued shroom boom. Bricks, packing material, leathers made of mushroom stuff, techno-scientific landscapes. Foods and medicine like lion's mane tinctures, mushroom bacon made from the asexual, non-mushrooming stage of oyster, koji-fermented everything, "superfood" supplementation in all forms. Anything that has the smack of novelty (though in some cases it's rather ancient technology rebranded as the latest discovery). Those who took my question to be about the future of myco-markets and heavily branded solutions to environmental degradation in the form of myco-remediation either hoped that some different people could also get a seat at the table to do some profiting (but were not holding their breath), or foresaw snake oil, charlatanism, and cheap tricks abounding until people stopped caring again, the waves of awe and wonder that people felt for fungi having been commodified into unenchanted dust. Others knew that the table in question is on fire, and that the flames are not the cleansing sort.

Because we are living the dystopia, where war is peace and the best some people can imagine is mushroom-as-savior, it makes so much sense to wonder why the future should be any different. Why shouldn't capitalism ruin mushrooms just as it does everything else? One respondent saw the inexorable approach of human-mushroom hybridity, through infection and technology, and was compelled by questions about how best to meet those moments in the now. Another wanted to leave it at "We're fucked, so let's document all the life we can." Maybe the only reasonable way to interpret the possibilities of the mycofuturistic is to decide how to navigate it as an extension of how we seem to be navigating the present. But it matters which present one has focused on. Another interviewee wondered about present efforts to map and chart the mycelium, what it will mean to have a census of these secretive beings, what attempts to exploit fungi could look like once certain powers know, so to speak, where they live. What is the balance between mapping in order to account for, to learn about, and, if need be, to protect–and the mapping that names a thing as a resource for future plunder? Whither, then, the mystery? Do fungi deserve their privacy? Another respondent wondered if fungi are even exploitable, thought that really, we would be playing ourselves if we believed we could impose a will on them that wasn't also theirs.

I'm eliding authorship here because each living thing, and the hyperobjects they collectively bring into being, will be the coauthors of an undescribed tomorrow. There are those who are adept at prophecy, the ones who have been handled with the least care and have therefore learned the art of prediction. But even the script lichen on the skin of a birch doesn't know what story it will end up writing, though its calligraphic flourishes do describe this space and time.

I can tell you what I want the mycofuture to look like: The mycelial networks of mutual aid. The maroon and Taíno, inventing jerk chicken, walking barefoot in the bush, foraging mushrooms. The endless web of couches and guest rooms available to us around the world. The de-emphasis of money. Reacquaintance with indeterminacy, with open expectations, with what is available versus what one wants. The windows into the smallest details of the forest floor, just about everywhere, shared just because. The care forged in Minneapolis fire. The unofficial belowground economies of the petit marronage. The anarchism that has always marked Black life. The wonder and reciprocity of truly living on the land we've been divorced from, which I glimpse in participants' faces when I lead mushroom walks. The things I want to see in the future are things that now survive mostly underground, but I want to see them fruiting everywhere.

Mario Ceballos may have felt any number of ways about the "mycofuture" when we spoke—I can't say for sure, since this Yaqui (Yoeme) luminary took the reins and told his story in an off-the-cuff, indeterminate way. But based on his approach and what he told me, I think it's fair to say he's at work on the myco-future as we speak.

I followed the organization he cofounded, POC Fungi Community (POCFC), on socials as soon as I became aware of the work they were doing out west, headquartered on Kumeyaay land, in so-called San Diego County. They remained on my radar as a collective of mycophiles putting together forays, workshops, and community events "working at the intersections of climate justice, food access, and social justice" to foster mushroom love, squadhood, and connection to the land among people of color.[9]

But I knew I had to link up with Mario and the crew beyond just a "follow" when Sneha shared a gorgeous piece of art she'd made, inspired by one of POCFC's "young elders, Lucian, who brilliantly renamed the 'false turkey tail,'" or *Stereum ostrea* and *S. hirsutum*, dubbing it the vulture tail. This was in the context of a foray in which the group had been discussing ways of knowing our kin, the mushrooms, and how, if you predicated your entire relationship with an organism on what you had hoped it would be, rather than actually seeing that organism for what they were, you were setting that relationship up for failure. As a name, "false turkey tail" clearly fell short. Lucian, eight at the time, supplied the name of "vulture tail," and as the group thought about the winged creature's work as a decomposer of the dead, the name came to seem more and more right. It was decided that this was the new name, then and there.

As soon as I read this brief anecdote, I was running to correct all the field guides in my possession with a permanent marker. I also use the name that Lucian coined when I lead walks. The story inspired Sneha's art piece, a collage with vulture tail fruit bodies, paper made with *Stereum* and several polypores, blood, turmeric, and wood charcoal. It won the North American Mycological Association's Visual Arts Contest that year.

It's no surprise that Mario's vision for community occasioned this invitation to heal a broken name. He's a born healer. While it was expected he would follow in his father's footsteps with construction work and "getting into cars," Mario gravitated toward the labor of care, working in hospitals and elder facilities, seeking out nursing roles and other jobs that were all about acting as a liaison and nurturing the unwell and vulnerable. Eventually, however, he became disenchanted with the hospital ecosystem and grew more interested in getting to the root of the illnesses afflicting those with his background. He took a leave of absence to look after his growing family and clear his head, and in the meantime embarked on research into preventative medicine, with a view to understanding what was ailing those who looked like him. This, combined with the whiteness of the spaces where he went to deepen his knowledge of plants and fungi, led to that intersection of social justice, climate justice, and food access, and planted the spore of POC Fungi Community. The one-year leave of absence from the hospital has lasted over a decade.

During our virtual chat, Mario gives me a tour of his altar, replete with beautiful mushroom art. I'm especially taken with glyphs depicting Xochipilli, the Aztec god of art and pleasure among other things, the way his image is run through with psilocybin mushrooms. Mario holds up to the screen an abalone shell with the kind of startling iridescence that hints at the celestial. These shells, he informs me, were portals for the Kumeyaay people to the spirit world. He shows me artwork depicting the fungus huitlacoche (*Ustilago maydis*) growing on beautiful multicolored gem corn. Huitlacoche is of special interest to Mario, and a focus of the work POCFC is engaged in at the time of this writing, because Mario believes this organism that is parasitic on corn (and frankly delicious in a quesadilla—I devoured several of the smoky delights while in Oaxaca) is not only doing important work ecologically, but is a major object lesson in how indigenous ways of knowing provide the key to living sustainably with the land.

Huitlacoche love is ancient mycophilia. The Aztecs celebrated the unexpected bounty when they saw it growing on their maize, and it's been a delicacy in Mexico ever since. Its appeal is growing elsewhere, but US farmers are still weirded out by the sight of the whitish, bluish, grayish galls that *Ustilago maydis* will occasionally bless them with. But it turns out that big agriculture may have gotten another thing wrong, y'all. In attempting to eradicate what it had the gall to christen "corn smut" with fungicide, the USDA may have tipped the scales further in the direction of catastrophic climate change. Mario tells me he has been watching with interest research that suggests the fungus does herculean labor in the soil sequestering carbon, while its incredibly prolific spore dispersal, as with the spores of other fungi, form a massive biomass that has the ability to make clouds and affect the weather. If this view is borne out, the notion that huitlacoche's world-shaping abilities have been stifled, out of a misguided sense that its visitation on the crop was anything other than lucky, is yet another tragic irony of wresting control of the land from those who know the most about how to care for it.

Mario takes me on a short virtual walk through his neighborhood. From my perch in Massachusetts, I'm treated to views of California chaparral woodland leading down to the valley, a field of very happy wildflowers that look to be chest high, the invasive eucalyptus that may be diminishing fungal populations with their natural fungicides, the palms that are so associated with Southern California but are not indigenous to those lands. Speaking with Mario made me want to get outside, organize a walk, call my grandmother, make more art, learn every single inch of this land or die trying. But it was the last thing he shared with me that made my eyes sting, so I had to look away from the screen.

Mario had attended the New Moon Mycology Summit in Vermont the year before. By all accounts, not just his, it was a transformative event. As a West Coaster in the wilds of Vermont and a lover of fox fire—the luminescent fungus *Panellus stipticus*—he was excited to be in the home range of one of the natural world's most brilliant showings. *P. stipticus* is a remarkable, smallish, ubiquitous, shellfish- or kidney-shaped, tan to brown mushroom. Its cap sometimes cracks where it has dried out and rehydrated multiple times as it ages—its stipe is generally short but well-defined. If one is really looking for them in my home woods after a generous rain, it's near impossible not to spot their overlapping clusters. Even during a drought, they're shriveled and hardened but always there—on a dry ID walk I'll look for a *P. s.* stick to hold aloft while rhapsodizing about the important role it and other white-rot fungi play in carbon sequestration. *P. s.* keeps them rapt. Also known as the bitter oysterling, the Latin name is descriptive of the astringent fruit bodies, which can staunch the flow of blood. Sometimes I'll collect several sticks covered in them and visibly shot through with their thick mycelial threads to make a weakly glowing terrarium—the first of this two-step process is to pop the sticks into an oversized mason jar. When I remember to water the setup, it will glow weakly from the corner of my room on those insomniac nights when my eyes are well-adjusted to lightlessness. The green glow is due to a biosynthetic cycle in which the fungus's adorably named luciferin and luciferases (themselves inexact monikers for less memorably named enzymes) do their thing and "yield an unstable high-energy intermediate that emits light as it becomes oxyluciferin"—in other words, they cast a heatless, eerie glow when the chemistry is chemistrying and conditions allow.[10]

As Mario described the nighttime excursion, I recalled one I'd participated in the previous July at a mushroom camp in New Hampshire. Participants held UV torches, so the trail was too bright for foxfire, but just bright enough that sphinx moths descended on our group like an enchantment. Earlier that day we'd found their large pupae alight with bright orange *Cordyceps militaris* clubs, so the moment felt pitched to remind us that to everything there is a season.

In my experience, you're likeliest to see that glow when the woods are uncomfortably wet, and it had been doing its fair share of raining in Vermont when Mario attended the summit. Another attendee said they had spied *P. stipticus* earlier in the weekend and offered to take him and eight others to go see if it was glowing at around midnight. Mario was excited. The Vermont trail was lit only by solar lights at distant intervals, so the going was very dark but not impossible. There was an actual factual meteor shower up above, and while a thunderstorm had just cleared the immediate vicinity, lightning could still be seen flashing above the canopy.

After a trek that was sometimes smooth, sometimes clumsy, and always thrilling, they arrived at the darkest part of the forest. The blackness was impenetrable save for the dance of the stars up above, the faint frozen sunshine of the ground lanterns, the company Mario was flanked by. They myceliated their way into the inky shadow, into the northern forest vividly damp but invisible around them. Up ahead there was another bright lantern and they moved toward it. As Mario spoke, I imagined everyone in the group, their pupils massively dilated, holding that brilliance at each pupil's center. This lantern looked different, wrong. The eight of them were quiet until the one who'd been leading piped up: "You guys know . . . you know that's the mushroom, right?"

This was met with jocular disbelief—of course there was no way biological luminescence was that bright. He had to be kidding. But as they moved closer, sure enough, it was fox fire.

A cluster of it: supple and fecund, intricate as a city, blazing bright as any man-made light, radiantly unexpected, the kind of beacon that draws kinfolk together, questing and questioning, just ahead if you trust in the journey, undescribed and indescribable, but we try, anyway, like a secret that's always there whether you look or not, so that you have to laugh to freight its finding lest the magic sweep you away in its lightness, and there again it *is*, on the horizon, waiting and not waiting, lighting the way.

NOTES

Introduction

1. George W. Hudler, *Magical Mushrooms, Mischievous Molds* (Princeton University Press, 1998), 26.

2. N. G. Ravichandra, *Fundamentals of Plant Pathology* (PHI Learning, 2013), 499.

3. Merlin Sheldrake, *Entangled Life* (Random House, 2020), 51.

4. A. Antonelli et al., *State of the World's Plants and Fungi 2023* (Royal Botanic Gardens, 2023), 12, 14.

5. Joel Hagen, "Five Kingdoms, More or Less: Robert Whittaker and the Broad Classification of Organisms," *Bioscience* 62, no. 1 (2012): 67–74.

6. Rachel Zoller (@yellowelanor), "Mushroom Shirt," Instagram, May 22, 2020, www.instagram.com/p/CAgl-gmhgac/.

7. Fiona Harvey, "World's Vast Networks of Underground Fungi to Be Mapped for First Time," *The Guardian*, Nov. 30, 2021.

In Search of Junjo (The Jamaican Mushroom)

1. Paul Farley and Michael Symmons Roberts, *Edgelands* (Vintage, 2012).

2. William Bryant Logan, *Dirt* (W. W. Norton, 2007), 20.

3. Alice Notley, *The Descent of Alette* (Penguin, 1992), 3.

4. Robert Wallace Thompson, "Mushrooms, Umbrellas, and Black Magic: A West Indian Linguistic Problem," *American Speech* 33, no. 3 (1958): 170–75.

5. M. Sandmann, "Un problema de geografía lingüística antillana," *Nueva Revista de Filología Hispánica* 9, no. 4 (1955): 383.

6. Patrick Browne, *The Civil and Natural History of Jamaica in Three Parts* (Osborne and Shipton, 1756), 78.

7. Thompson, "Mushrooms," 171.

8. Thompson, "Mushrooms," 175.

9. Frederic G. Cassidy, "Some Footnotes on the 'Junjo' Question," *American Speech* 36, no. 2 (1961): 102.

10. B. W. Higman, "Jamaican Versions of Callaloo," *Callaloo* 30, no. 1 (2007): 351–68.

11. Higman, "Jamaican Versions," 364.

12. Higman, "Jamaican Versions," 364.

13. Gary Lincoff, *The Complete Mushroom Hunter* (Quarto Books, 2017), 12.

14. Malaramuthan R., "Once poor man's food, mushrooms elevated to five-star dish," *Inmathi* (Chennai, India), Nov. 26, 2022, https://inmathi.com/2022/11/26/once-poor-mans-food-mushrooms-elevated-to-five-star-dish/71766.

15. Tao Leigh Goffe, "Kitchen Marronage: A Genealogy of Jerk," *The Funambulist* 31 (2022), https://thefunambulist.net/magazine/politics-of-food/kitchen-marronage-a-genealogy-of-jerk-tao-leigh-goffe.

16. Daphne Ewing-Chow, "Sovereignty and the Soil: Chief Richard Currie and the Rise of the Maroon Nation in Jamaica," *Forbes*, Feb. 28, 2021.

17. William Delisle Hay, *An Elementary Text-Book of British Fungi* (London, 1887), 6–7.

18. For the Ghanian tale, see W. H. Barker and Cecilia Sinclair, *West African Folk-Tales* (Harrap, 1917), 177–81; for the Ugandan story, Damascus Kafumbe, *Tuning the Kingdom* (University of Rochester Press, 2018), 44; for the Aka song, Bayaka Women, "Women Gathering Mushrooms," recorded by Louis Sarno and Bernie Krause ca. 1985, track 1 on *The Music of the Bayaka*, Wild Sanctuary, 2007, CD; on transmission of mushroom knowledge among Zimbabwean women, see *100 Women: The Mushroom Woman*, presented by Chido Govera (BBC Documentary, 2020), www.bbc.co.uk/sounds/play/w3ct1cs1.

19. Kafumbe, *Tuning the Kingdom*, 44.

20. David Cappaert et al., "Emerald Ash Borer in North America: A Research and Regulatory Challenge," *American Entomologist* 51, no. 3 (2005): 152–65.

21. Massachusetts Department of Conservation and Recreation, "Emerald Ash Borer in Massachusetts," Mass.gov, accessed May 30, 2024, www.mass.gov/guides/emerald-ash-borer-in-massachusetts.

22. John T. Edge, "The Hidden Radicalism of Southern Food," *New York Times*, May 6, 2017.

23. Renard, Cerberus, and MangoSweet, comments on Joseph, "Mushroom on Segment 11 of Waitukubuli Nature Trail," *Dominica News Online*, March 21–22, 2011, https://dominicanewsonline.com/news/homepage/features/photo-of-the-day/mushroom-on-segment-11-of-waitukubuli-nature-trail.

24. Adam Levy, "Hurricane Maria Rain Amount Chances Are Boosted by Climate Change," *Scientific American*, April 26, 2019, www.scientificamerican.com/podcast/episode/hurricane-maria-rain-amount-chances-are-boosted-by-climate-change/.

25. "World Directory of Minorities and Indigenous Peoples—Dominica," refworld, Minority Rights Group International, 2007, https://webarchive

.archive.unhcr.org/20230518080030/https://www.refworld.org/docid/4954ce32c.html.

26. On the "poison arrow curtain," see Troy S. Floyd, *The Columbus Dynasty in the Caribbean, 1492–1526* (University of New Mexico Press, 1973), 135.

27. Lauren Ré, "Mycology in the Tropics: Fungi of Guyana & Trinidad," South Sound Mushroom Club, Mar. 12, 2023, video, 1:16, https://youtu.be/RIiORa54FzI?si=cU2PAlVM4X2wykvM.

28. Eric Garraway, "In Search of Jamaica's Gardening Ant," Natural History Society of Jamaica, accessed May 30, 2024, http://naturalhistorysocietyjamaica.org/arthropods/JamaicasGardeningAnt_EGarraway.pdf.

29. "Clans and Totems in Buganda Culture," BeingAfrican, accessed May 30, 2024, https://beingafrican.com/clans-and-totems-in-buganda (page discontinued).

In Search of *Tuber melanosporum* (The Black Winter Truffle)

1. Ian R. Hall et al., *Taming the Truffle* (Timber Press, 2007), 32.

2. Federico Kukso, "How Truffles Took Root Around the World," *Smithsonian Magazine*, Oct. 31, 2022.

3. Ahmed Mustafa et al., "An Overview on Truffle Aroma and Main Compounds," *Molecules* 25, no. 24 (2020): 5948.

4. On anandamide, see Rachel Nuwer, "Truffles Have a THC-Like Substance in Them," *Smithsonian Magazine*, Dec. 22, 2014, www.smithsonianmag.com/smart-news/truffles-have-thc-substance-them-180953705/. Laura Cullere et al., "Characterisation of Aroma Active Compounds in Black Truffles (*Tuber melanosporum*) and Summer Truffles (*Tuber aestivum*) by Gas Chromatography-Olfactometry," *Food Chemistry* 122, no. 1 (2010): 300.

5. Hall et al., *Taming the Truffle*, 33.

6. Rowan Jacobsen, *Truffle Hound* (Bloomsbury, 2021), 23.

7. Jason Horowitz, "Hunting for Truffles Is a Perilous Pursuit, Especially for the Dogs Who Dig," *New York Times*, updated Feb. 14, 2023.

8. "French Farmer Jailed for Killing Suspected Truffle Thief," *The Guardian*, May 29, 2015, www.theguardian.com/world/2015/may/29/french-farmer-jailed-for-killing-suspected-truffle-thief.

9. Giovanni Pacioni et al., "Truffles Contain Endocannabinoid Metabolic Enzymes and Anandamide," *Phytochemistry* 110 (2015): 104–110, https://pubmed.ncbi.nlm.nih.gov/25433633/.

10. For a discussion of how many odors the human mind can recall, see Angelika Börsch, "Small Molecules Make Scents," *Science in School*, no. 6 (2007), www.scienceinschool.org/article/2007/scents.

11. Albert Keim and Louis Lumet, *Honoré de Balzac* (Frederick A. Stokes Company, 1914), 205.

12. Graham Robb, "The Pens of the Musketeer," *New York Review of Books*, March 20, 2008.

13. Joel Dreyfuss, "Gerard Depardieu Plays Dumas amid Some Controversy," *Star-Tribune* (Minneapolis, MN), March 19, 2010.

14. Lawrence Millman, *Fungipedia* (Princeton University Press, 2019), 140.

15. The definition of "tuckahoe" is from Millman, *Fungipedia*, 140.

In Search of *Ophiocordyceps unilateralis* (The Zombie-Ant Fungus)

1. Mike Mariani, "The Tragic, Forgotten History of Zombies," *The Atlantic*, Oct. 28, 2015, www.theatlantic.com/entertainment/archive/2015/10/how-america-erased-the-tragic-history-of-the-zombie/412264/.

2. Zora Neale Hurston, *Tell My Horse* (1938; Harper Perennial, 2009), 179.

3. Hurston, *Tell My Horse*, 183.

4. Richard Dawkins, *The Extended Phenotype* (Oxford University Press, 1989), xiii.

5. Charissa de Bekker and Biplabendu Das, "Hijacking Time: How Ophiocordyceps Fungi Could Be Using Ant Host Clocks to Manipulate Behavior," *Parasite Immunology* 44, no. 3 (2022): e12909.

6. Keiichi Nakahara et al., "Gut Microbiota of Parkinson's Disease in an Appendectomy Cohort: A Preliminary Study," *Scientific Reports* 13, no. 1 (2023): 2210.

7. Westin Porter, "Zombie Ants," Natural History Museum of Utah, March 6, 2018, https://nhmu.utah.edu/articles/2023/05/zombie-ants.

8. David Pacchioli, "Getting to the Bottom of the Zombie Ant Phenomenon," Penn State University, May 21, 2013, www.psu.edu/news/research/story/getting-bottom-zombie-ant-phenomenon/.

9. "Health Topics–Suicide Prevention," Polaris, last reviewed July 19, 2022, www.cdc.gov/policy/polaris/healthtopics/suicide/index.html (page discontinued).

10. Lee Brown, "Get Ready for Sex-Crazed Zombie Cicadas Known as 'Flying Saltshakers of Death,'" *New York Post*, May 20, 2021, https://nypost.com/2021/05/20/get-ready-for-sex-crazed-zombie-cicadas-in-new-york; Ed Yong, "How the Zombie Fungus Takes Over Ants' Bodies to Control Their Minds," *The Atlantic*, Nov. 14, 2017, www.theatlantic.com/science/archive/2017/11/how-the-zombie-fungus-takes-over-ants-bodies-to-control-their-minds/545864/.

11. Mariani, "Zombies."

12. *The Girl with All the Gifts*, directed by Colm McCarthy (Warner Bros. Pictures, 2016).

13. M. R. Carey, *The Girl with All the Gifts* (Orbit Books, 2014), 1.

14. Carey, *Girl with All the Gifts*, 40.

15. Maisy Flowers, "Night of the Living Dead: Why George Romero Rewrote Ben's Character," *Screen Rant*, June 20, 2020, https://screenrant.com/night-living-dead-george-romero-rewrite-ben-character-duane-jones-reason.

16. Carey, *Girl with All the Gifts*, 1.

17. Matthew Soules, *Icebergs, Zombies, and the Ultra-Thin* (Princeton Architectural Press, 2021).

18. Andy Newman and Michael Gold, "New York City Clears 239 Homeless Camps," *New York Times*, March 30, 2022.

19. "Basic Facts About Homelessness: New York City," Coalition for the Homeless, updated May 2024, www.coalitionforthehomeless.org/basic-facts-about-homelessness-new-york-city.

20. Geoffrey Propheter, "From Tax Breaks to Affordable Housing: Examining the 421-a Tax Exemption for One57," New York City Independent Budget Office, July 2015, www.ibo.nyc.ny.us/iboreports/from-tax-breaks-to-affordable-housing-examining-the-421-a-tax-exemption-for-one57-july-15-2015.pdf.

21. Jacob L. Bender, *Modern Death in Irish and Latin American Literature* (Springer International Publishing, 2020), 176–77.

22. Bender, *Modern Death*, 177.

23. Alejo Carpentier, *The Kingdom of This World* (Farrar, Straus and Giroux, 1989), 178.

In Search of *Amanita phalloides* (The Death Cap)

1. On "phármakon," see Jacques Derrida, "Plato's Pharmacy," in *Dissemination*, trans. Barbara Johnson (University of Chicago Press, 1981), 70. On "pharmakós," see Jan Bremmer, "Scapegoat Rituals in Ancient Greece," *Harvard Studies in Classical Philology* 87 (1983): 300.

2. John Rashford, "Those That Do Not Smile Will Kill Me: The Ethnobotany of the Ackee in Jamaica," *Economic Botany* 55, no. 2 (2001): 190–211.

3. "Mushroom Poisoning Syndromes," North American Mycological Association, accessed May 30, 2024, https://namyco.org/interests/toxicology/mushroom-poisoning-syndromes.

4. Fez Inkwright, *Botanical Curses and Poisons* (Sterling Ethos, 2021), 115.

5. The quotation about the West African slave ship is from Jose Jackson-Malete et al., "Natural Toxins in Fruits and Vegetables: *Blighia sapida* and

Hypoglycin" in *Food Safety and Quality Systems in Developing Countries*, vol. 1, *Export Challenges and Implementation Strategies*, ed. André Gordon (Elsevier, 2015).

6. On the history of uses of ackee before the Atlantic slave trade, see André Gordon and Jose Jackson-Malete, "The Life Cycle of Ackee (*Blighia sapida*): Environmental and Other Influences on Toxicity" in *Food Safety and Quality Systems in Developing Countries*, vol. 1, *Export Challenges and Implementation Strategies*, ed. André Gordon (Elsevier, 2015). See also Brendan Sainsbury, "Ackee and Saltfish: Jamaica's Breakfast of Champions," BBC, March 15, 2021, www.bbc.com/travel/article/20210315-ackee-and-saltfish-jamaicas-breakfast-of-champions.

7. Alejo Carpentier, *The Kingdom of This World* (Farrar, Straus and Giroux, 1989), 23–36.

8. Jamaica Kincaid, *Annie John* (Farrar, Straus and Giroux, 1998), 78.

9. Hugh Thomas, *The Slave Trade* (Simon and Schuster, 1997), 197.

10. William Henry Whitmore, *The Massachusetts Civil List for the Colonial and Provincial Periods, 1630–1774* (Albany, NY: J. Munsell, 1870), 11.

11. Elise Lemire, *Black Walden* (University of Pennsylvania Press, 2009), 43–59.

12. On Mark's body, see Lemire, *Black Walden*, 59. Paul Revere to Jeremy Belknap, circa 1798, Massachusetts Historical Society, www.masshist.org/database/viewer.php?item_id=99.

13. Sasha Turner Bryson, "The Art of Power: Poison and Obeah Accusations and the Struggle for Dominance and Survival in Jamaica's Slave Society," *Caribbean Studies* 41, no. 2 (2013), 62–63, www.redalyc.org/pdf/392/39230911003.pdf.

In Search of *Psilocybe* Species (The Magic Mushroom)

1. Breyten van der Merwe et al., "A Description of Two Novel *Psilocybe* Species from Southern Africa and Some Notes on African Traditional Hallucinogenic Mushroom Use," *Mycologia* 116, no. 5 (2024): 821–34.

2. Stephen Kelly, "'Silicon Valley': A Roast or Real Life?," *Wired*, Jan. 6, 2015, www.wired.com/story/silicon-valley.

3. *Silicon Valley*, season 1, episode 3, "Articles of Incorporation," directed by Tricia Brock, written by Matteo Borghese, Rob Turbovsky, and Mike Judge, featuring Thomas Middleditch and T. J. Miller. Aired April 20, 2014. HBO Entertainment.

4. "Ancient Medicine with Modern Application," MycoMeditations, accessed May 30, 2024, www.mycomeditations.com/careers.

5. Carlene Davis, "'High'-Risk Retreats: Hundreds of Americans Swarm JA for Mushroom Drug Sessions," *The Gleaner* (Kingston, Jamaica), Jan. 15,

2020, https://jamaica-gleaner.com/article/lead-stories/20200115/high-risk-retreats-hundreds-americans-swarm-ja-mushroom-drug-sessions.

6. Eric Mr. Myco, "Concerning New Trend," Treasure Beach Forum, May 6, 2015, www.treasurebeach.net/forum_static/webtonic.info/discus/messages/1584/24321f585.html?1433021204.

7. C. A. Ruck et al., "Entheogens," *Journal of Psychedelic Drugs* 11, nos. 1–2 (1979), 145–46, https://doi.org/10.1080/02791072.1979.10472098.

8. On per capita GDP in Jamaica, see United States Central Intelligence Agency and National Foreign Assessment Center, "Jamaica," *The World Factbook*, updated May 28, 2024, www.cia.gov/the-world-factbook/countries/jamaica/#economy.

9. Chloe Aridjis, "On María Sabina, One of Mexico's Greatest Poets," British Council, March 30, 2015, www.britishcouncil.org/voices-magazine/maria-sabina-one-of-mexicos-greatest-poets.

10. Logan Neitzke-Spruill, "Race as a Component of Set and Setting: How Experiences of Race Can Influence Psychedelic Experiences," *Journal of Psychedelic Studies* 4, no. 1 (2019): 51–60, https://doi.org/10.1556/2054.2019.022.

11. Marcia Elizabeth Sutherland, "Toward a Caribbean Psychology: An African-Centered Approach," *Journal of Black Studies* 42, no. 8 (2011): 1176, https://doi.org/10.1177/0021934711410547.

12. Simon Yugler, "Anima Mundi: Psychedelics and the Ensouled Earth," *Psychedelics Today*, Oct. 6, 2021, https://psychedelicstoday.com/2021/10/06/anima-mundi-psychedelics-and-the-ensouled-earth.

13. José Oliver, "Taino Ritual Seat," *A History of the World in 100 Objects*, BBC, accessed May 30, 2024, www.bbc.co.uk/ahistoryoftheworld/objects/nknww6EoQO-nKTncBvhNbw.

14. Colorado Springs City Council, "City Council Meeting 08_09_22," Facebook, August 9, 2022, www.facebook.com/watch/live/?ref=watch_permalink&v=364296622574935.

15. Tiney Ricciardi, "What We Know About the Money Behind Psychedelic Mushroom Legalization in Colorado," *Denver Post*, Nov. 3, 2022, www.denverpost.com/2022/11/03/proposition-122-psilocybin-legalization-new-approach-pac-donors.

16. Robert Frost, "Birches," *The Robert Frost Reader* (Henry Holt, 2002), 63.

In Search of *Fomes excavatus* (The Tinder Fungus)

1. Isaiah 66:15 (King James Version).

2. Ellie Shechet, "This Fire-Loving Fungus Eats Charcoal, If It Must," *New York Times*, Nov. 28, 2021, www.nytimes.com/2021/11/28/science/fungus-wildfire-charcoal.html.

3. Alex Dorr, "Fire Loving Fungi," *Mushroom Revival* (podcast), February 16, 2022, accessed May 30, 2024, www.mushroomrevival.com/blogs /podcast/fire-loving-fungi.

4. Keewaydinoquay, *Puhpohwee for the People: A Narrative Account of Some Uses of Fungi among the Ahnishinaubeg* (Botanical Museum of Harvard University, 1978), 1–3; Marian Berihuete-Azorín et al., "Punk's Not Dead: Fungi for Tinder at the Neolithic Site of La Draga (NE Iberia)," *PLoS One* (April 25, 2018): 12–13, https://doi.org/10.1371/journal.pone.0195846.

5. Ian R. Hall et al., *Taming the Truffle* (Timber Press, 2007), 167–171.

6. Radames J. B. Cordero and Arturo Casadevall, "Functions of Fungal Melanin Beyond Virulence," *Fungal Biology Reviews* 31, no. 2 (2017): 99, https://doi.org/10.1016/j.fbr.2016.12.003.

7. David Moore et al., "Gravimorphogenesis in Agarics," *Mycological Research* 100, no. 3 (1996), 257, https://doi.org/10.1016/S0953-7562(96)80152-3.

8. Stephen J. Pyne, *Fire: A Brief History* (University of Washington Press, 2019), 8.

9. On the term "Plantationocene," see Donna Haraway et al., "Anthropologists Are Talking—About the Anthropocene," *Ethnos* 81, no. 3 (2016): 535–64, https://doi.org/10.1080/00141844.2015.1105838.

10. "Ötzi's Clothing," Ötzi the Iceman, South Tyrol Museum of Archaeology, accessed May 30, 2024, www.iceman.it/en/clothing/; "The Iceman's Equipment," Ötzi the Iceman, South Tyrol Museum of Archaeology, accessed May 30, 2024, www.iceman.it/en/equipment/.

11. M. Fabiola Pulido-Chavez et al., "Rapid Bacterial and Fungal Successional Dynamics in First Year After Chaparral Wildfire," *Molecular Ecology* 32, no. 7 (2023): 1685–1707, https://doi.org/10.1111/mec.16835.

12. Monika S. Fischer et al., "Pyrolyzed Substrates Induce Aromatic Compound Metabolism in the Post-Fire Fungus, *Pyronema domesticum*," *Frontiers in Microbiology* 12 (2021), https://doi.org/10.3389/fmicb.2021 .729289.

13. Kate Mather, "Hunter Charged with Sparking Massive Rim Fire, State's Third-Largest," *Los Angeles Times*, Aug. 7, 2014, www.latimes.com /local/lanow/la-me-ln-rim-fire-charges-20140807-story.html.

14. Thomas D. Bruns et al., "A Simple Pyrocosm for Studying Soil Microbial Response to Fire Reveals a Rapid, Massive Response by *Pyronema* Species," *PloS One* 15, no. 3 (2020): e0222691, https://doi.org/10.1371 /journal.pone.0222691.

15. Rebecca Etter, "A Timeline of Events Surrounding 'Cop City' and the Dekalb/Blackhall Land Swap," WABE (Atlanta, GA), January 31, 2023, www.wabe.org/a-timeline-of-events-surrounding-cop-city-and-the-dekalb -blackhall-land-swap/.

16. Merlin Sheldrake, *Entangled Life* (Random House, 2020), 50.

17. R. J. Rico, "Muddy Clothes? 'Cop City' Activists Question Police Evidence," Associated Press, March 23, 2023, https://apnews.com/article/cop-city-protest-domestic-terrorism-atlanta-6d114e109d489d316f588f51c7cabocc; Sean Keenan and Rick Rojas, "'Cop City' Prosecutions Hinge on a New Definition of Domestic Terrorism," *New York Times*, Feb. 27, 2024, www.nytimes.com/2024/02/26/us/cop-city-domestic-terrorism.html.

18. Keenan and Rojas, "Domestic Terrorism," 2024.

19. Odette Yousef, "RICO Case Against Cop City Protesters in Atlanta Stirs Concerns About Free Speech," NPR, Sept. 21, 2023, www.npr.org/2023/09/21/1200898062/rico-case-against-cop-city-protesters-in-atlanta-stirs-concerns-about-free-speec.

20. Christopher E. Bruce and Hina Shamsi, "RICO and Domestic Terrorism Charges Against Cop City Activists Send a Chilling Message," ACLU, Sept. 21, 2023, www.aclu.org/news/free-speech/rico-and-domestic-terrorism-charges-against-cop-city-activists-send-a-chilling-message.

21. "Anarchism Must Not Be Criminalized: Statement from 12 of the 61 People Indicted on RICO Charges in Atlanta," Institute for Anarchist Studies, Nov. 20, 2023, https://anarchiststudies.org/anarchism-must-not-be-criminalized/.

22. E.g., June Hidalgo et al., "Mycoremediation with *Agaricus bisporus* and *Pleurotus ostreatus* Growth Substrates Versus Phytoremediation with *Festuca rubra* and *Brassica* Sp. for the Recovery of a Pb and γ-HCH Contaminated Soil," *Chemosphere* 327 (2023): 138538, https://doi.org/10.1016/j.chemosphere.2023.138538.

23. Sean Summers, "'Tortuguita Vive': A No-Compromise Movement Responds to Police Killing a Forest Defender," *Unicorn Riot*, Feb. 27, 2023, https://unicornriot.ninja/2023/tortuguita-vive-a-no-compromise-movement-responds-to-police-killing-a-forest-defender/.

24. Rosana Hughes and Caroline Silva, "Autopsy: Gunshot Residue 'Not Seen' on Activist Killed at Police Training Center," *Atlanta Journal-Constitution*, April 20, 2023, www.ajc.com/news/crime/activist-killed-at-police-training-center-site-had-more-than-50-gunshot-wounds-autopsy-finds/RSL5T2D7WJDXNIULLBL6ZUZADE/.

25. "Candlelight Vigil–*Multiclavula mucida*," *National Audubon Society Mushrooms of North America*, ed. Jim Cirigliano (Alfred A. Knopf, 2023), 94.

26. "Appalachian Basin Geology," John A. Dutton e-Education Institute at Pennsylvania State University, accessed May 30, 2024, www.e-education.psu.edu/earth109/node/973.

27. John A. Stanturf et al., "Fire in Southern Forest Landscapes," in *Southern Forest Resource Assessment–General Technical Report SRS-53*, eds. David

Wear and John Greis (Asheville, NC: US Department of Agriculture, Forest Service, Southern Research Station, 2002), accessed May 30, 2024, https://web.archive.org/web/20160808000609/http://www.srs.fs.fed.us/sustain/report/fire/fire.htm.

28. "Lightning," *National Geographic*, accessed May 30, 2024, www.nationalgeographic.com/environment/article/lightning.

29. Pyne, *Fire: A Brief History*, 8.

30. Chris Iorfida, "Self-Immolation as Political Protest: A Brief Modern History," CBC News, Feb. 28, 2024, www.cbc.ca/news/world/bushnell-self-immolation-deaths-1.7126655.

31. "Haitian Burns Self in Front of Mass. Capitol," *Los Angeles Times*, Aug. 31, 1987, www.latimes.com/archives/la-xpm-1987-08-31-mn-3348-story.html.

32. Jon Coburn, "Self-Immolation: Hundreds of People in the US Have Set Themselves on Fire in Protest Since the 1960s," *The Conversation*, April 25, 2024, https://theconversation.com/self-immolation-hundreds-of-people-in-the-us-have-set-themselves-on-fire-in-protest-since-the-1960s-228438.

33. Ellie Silverman and Ian Shapira, "Outside the Supreme Court, a Life of Purpose and Pain Ends in Flames," *Washington Post*, April 26, 2022, www.washingtonpost.com/dc-md-va/2022/04/26/wynn-bruce-fire-supreme-court-climate-activist/.

34. "Protester Critically Injured After Setting Self on Fire Outside Israeli Consulate in Atlanta," Associated Press, Dec. 2, 2023, https://apnews.com/article/israeli-consulate-self-immolation-atlanta-protester-8f17dd72592f86797a45cda9b60605a5.

35. Victoria Audu, "The Back End of Genocide," *The Republic*, Nov. 19, 2023, https://republic.com.ng/october-november-2023/congo-cobalt-genocide/.

36. George Sweeney, "Self-Immolative Martyrdom: Explaining the Irish Hungerstrike Tradition," *Studies: An Irish Quarterly Review* 93, no. 371 (2004): 343, www.jstor.org/stable/30095978.

37. Masha Gessen, "Aaron Bushnell's Act of Political Despair," *New Yorker*, Feb. 28, 2024, www.newyorker.com/news/our-columnists/aaron-bushnells-act-of-political-despair.

In Search of *Tricholoma magnivelare* (The Mushroom at the End of the World)

1. Anna Tsing, *The Mushroom at the End of the World* (Princeton University Press, 2015), 20, 27.

2. “Safe Harbor/AIDS Archive Collection: Overview,” Provincetown History Preservation Project, accessed May 30, 2024, https://provincetown historyproject.org/browse/view?browseCollection=23.

3. Bob Seay, “Here’s Where in Massachusetts the Pilgrims First Landed in 1620—And It Wasn’t Plymouth,” GBH, Nov. 11, 2020, www.wgbh.org /news/local/2020-11-11/heres-where-in-massachusetts-the-pilgrims-first -landed-in-1620-and-it-wasnt-plymouth.

4. Tsing, *Mushroom at the End of the World*, 79.

5. *Foragers*, directed by Jumana Manna (2022; Golan Heights, Galilee, Jerusalem), digital video, www.jumanamanna.com/Foragers.

6. S. Young et al., “A Meta-Analysis of the Prevalence of Attention Deficit Hyperactivity Disorder in Incarcerated Populations,” *Psychological Medicine* 45, no. 2 (2015): 247, https://doi.org/10.1017/S0033291714000762.

7. The phrase “summer’s easy riches” is a quotation from Tsing, *Mushroom at the End of the World*, 2.

8. Yasuyuki Aoki, “‘Matsutake’ Mushrooms Auctioned Off for 930,000 Yen,” *Asahi Shimbun*, Oct. 12, 2023, www.asahi.com/ajw/articles/15027006.

9. Jamaica Kincaid, *A Small Place* (Penguin, 1988), 6.

10. Faustino Hernández-Santiago et al., “Pictographic Representation of the First Dawn and Its Association with Entheogenic Mushrooms in a 16th-Century Mixtec Mesoamerican Codex,” *Revista Mexicana de Micología* 46 (2017): 19–28.

11. On the effect of drought on el Árbol del Tule, see Zsolt Debreczy and István Rácz, “El Árbol del Tule: The Ancient Giant of Oaxaca,” *Arnoldia* 57 (1997): 3–11.

In Search of *Schizophyllum commune* (The Charismatic Mushroom)

1. Andrew Adamatzky, “Language of Fungi Derived from Their Electrical Spiking Activity,” *Royal Society Open Science* 9, no. 4 (2022), https://doi .org/10.1098/rsos.211926.

2. Linda Geddes, “Mushrooms Communicate with Each Other Using up to 50 ‘Words,’ Scientist Claims,” *The Guardian*, April 5, 2022, www .theguardian.com/science/2022/apr/06/fungi-electrical-impulses -human-language-study.

3. Ellen Phiddian, “Newly Discovered: Smallest Fanged Frog in the World,” *Cosmos*, Dec. 21, 2023, https://cosmosmagazine.com/nature /animals/fanged-frog-smallest.

4. I am indebted here and throughout this chapter to Patricia Kaishian and Hasmik Djoulakian, “The Science Underground: Mycology as a Queer

Discipline," *Catalyst* 6, no. 2 (2020), https://doi.org/10.28968/cftt.v6i2.33523.

5. Eun Jeong Won et al., "Molecular Identification of Schizophyllum commune as a Cause of Allergic Fungal Sinusitis," *Annals of Laboratory Medicine* 32 (2012): 375–79, https://doi.org/10.3343/alm.2012.32.5.375; Vivian Tullio et al., "Schizophyllum Commune: An Unusual of Agent Bronchopneumonia in an Immunocompromised Patient," *Medical Mycology* 46, no. 7 (2008): 735–38, https://doi.org/10.1080/13693780802256091.

6. Aleksandra Bezmenova et al., "Rapid Accumulation of Mutations in Growing Mycelia of a Hypervariable Fungus Schizophyllum commune," *Molecular Biology and Evolution* 37, no. 8 (2020): 2279, https://doi.org/10.1093/molbev/msaa083.

7. David Hibbett, "Mycologist Answers Mushroom Questions from Twitter," *Wired*, May 2, 2023, www.wired.com/video/watch/tech-support-mycologist-answers-mushroom-questions-from-twitter.

8. Nawal El Saadawi interviewed by Krishnan Guru-Murthy, "Nawal El Saadawi on Feminism, Fiction, and the Illusion of Democracy," *Ways to Change the World*, Channel 4 News, video, www.youtube.com/watch?v=djMfFU7DIB8.

9. Justine Karst, Melanie D. Jones, and Jason D. Hoeksema, "Positive Citation Bias and Overinterpreted Results Lead to Misinformation on Common Mycorrhizal Networks in Forests," *Nature Ecology & Evolution* 7, no. 4 (2023): 501–11, https://doi.org/10.1038/s41559-023-01986-1.

10. Bev Betkowski, "Do Trees Really 'Talk' to Each Other Through Underground Fungal Networks?," University of Alberta Folio, Feb. 13, 2023, www.ualberta.ca/folio/2023/02/do-trees-really-talk-to-each-other-through-underground-fungal-networks.html.

11. Gabriel Popkin, "The Wood-Wide Web—A Story Too Good for Its Own Good?," *Gabriel Popkin* (blog), Nov. 23, 2023, www.gabrielpopkin.com/blog.

12. Tathagata Som and Kit Dobson, "'I Struggled with Anthropomorphisms': On the Problem of Metaphors, Happiness, and Forests in Finding the Mother Tree," *ISLE: Interdisciplinary Studies in Literature and Environment* (2024): isae048.

13. Imane Laraba et al., "Pseudoflowers Produced by *Fusarium Xyrophilum* on Yellow-Eyed Grass (*Xyris* Spp.) in Guyana: A Novel Floral Mimicry System?," *Fungal Genetics and Biology* 144 (2020): 103466, https://doi.org/10.1016/j.fgb.2020.103466.

14. Ching-Han Lee et al., "A Carnivorous Mushroom Paralyzes and Kills Nematodes via a Volatile Ketone," *Science Advances* 9, no. 3 (2023): eade4809, https://doi.org/10.1126/sciadv.ade4809.

15. Carolyn Merchant, *The Death of Nature: Women, Ecology, and the Scientific Revolution* (Harper & Row, 1989).

16. Bathsheba Demuth, "Do Whales Judge Us? Interspecies History and Ethics," lecture, Environment Forum at the Mahindra Center for the Humanities at Harvard University, Cambridge, MA, Dec. 5, 2019, https://youtu.be/yh_kJAoNaug?si=ZFx7W2uuEG3DtM93.

17. Carl Safina, "Why Anthropomorphism Helps Us Understand Animals' Behavior," *Medium* (blog), Aug. 22, 2016, https://medium.com/@carlsafina/why-anthropomorphism-helps-us-understand-animals-behavior-155b69535d17.

18. Robin Wall Kimmerer, *Braiding Sweetgrass* (Milkweed Editions, 2013), 56.

In Search of Undescribed Species (All We Don't Know)

1. Thomas Hariot, *A Briefe and True Report of the New Found Land of Virginia* . . . , trans. Richard Hakluyt (ca. 1590; New York: J. Sabin and Sons, 1871), https://docsouth.unc.edu/nc/hariot/hariot.html.

2. Eberhard August Wilhelm von Zimmermann, *Geographische Geschichte des Menschen und der allgemein verbreiteten vierfüssigen Thiere* (Leipzig: In der Weygandschen buchhandlung, 1778), https://archive.org/details/geographischegeoozimmgoog.

3. Gordon Whittington, "How Did Whitetail Deer Get Their Name? Part I," North American Whitetail, Feb. 1, 2022, www.northamericanwhitetail.com/editorial/how-did-whitetail-deer-get-their-name-part-1/456646.

4. James R. Heffelfinger, "Taxonomy, Evolutionary History, and Distribution" in *Biology and Management of White-Tailed Deer*, ed. David G. Hewitt (CRC Press, 2011), 5–6.

5. See Jessie Little Doe Fermino, "An Introduction to Wampanoag Grammar" (master's thesis, MIT, 2000), 13, https://core.ac.uk/download/pdf/4408498.pdf.

6. "Deer, Ruminant," Comparative Siouan Dictionary, compiled by Richard T. Carter et al., accessed May 30, 2024, https://csd.clld.org/parameters/922#4/41.64/-96.87.

7. Sara Maroske and Tom W. May, "Naming Names: The First Women Taxonomists in Mycology," *Studies in Mycology* 89, no. 1 (2018): 70–72, https://doi.org/10.1016/j.simyco.2017.12.001.

8. "In U.S., Decline of Christianity Continues at Rapid Pace," Pew Research Center, Oct. 17, 2019, www.pewresearch.org/religion/2019/10/17/in-u-s-decline-of-christianity-continues-at-rapid-pace/.

9. Elle Mari, "POC Fungi Community," Partner Spotlight, UC San Diego Center for Community Health, Nov. 17, 2023, https://ucsdcommunityhealth.org/about/partner-spotlight/blog/poc-fungi-community.

10. The quote is from Huei-Mien Ke and Isheng Jason Tsai, "Understanding and Using Fungal Bioluminescence: Recent Progress and Future Perspectives," *Current Opinion in Green and Sustainable Chemistry* 33 (2022), www.sciencedirect.com/science/article/pii/S2452223621001267.